# FACILITATING INTERDISCIPLINARY RESEARCH

Committee on Facilitating Interdisciplinary Research
Committee on Science, Engineering, and Public Policy

NATIONAL ACADEMY OF SCIENCES,
NATIONAL ACADEMY OF ENGINEERING, *AND*
INSTITUTE OF MEDICINE
*OF THE NATIONAL ACADEMIES*

THE NATIONAL ACADEMIES PRESS
WASHINGTON, D.C.
**www.nap.edu**

THE NATIONAL ACADEMIES PRESS   500 Fifth Street, N.W.   Washington, D.C. 20001

NOTICE: The project that is the subject of this report was approved by the Governing Board of the National Research Council, whose members are drawn from the councils of the National Academy of Sciences, the National Academy of Engineering, and the Institute of Medicine. The members of the committee responsible for the report were chosen for their special competences and with regard for appropriate balance.

Support for this project was provided by the W. M. Keck Foundation. Any opinions, findings, conclusions, or recommendations expressed in this publication are those of the author(s) and do not necessarily reflect the views of the organizations or agencies that provided support for the project.

International Standard Book Number 0-309-09435-6 (Book)
International Standard Book Number 0-309-54727-X (PDF)

Available from the Committee on Science, Engineering, and Public Policy, 500 Fifth Street, NW, Washington, D.C. 20001; 202-334-2807; Internet, http://www.nationalacademies.org/cosepup

Additional copies of this report are available from the National Academies Press, 500 Fifth Street, N.W., Lockbox 285, Washington, D.C. 20055; (800) 624-6242 or (202) 334-3313 (in the Washington metropolitan area); Internet, http://www.nap.edu

Grateful acknowledgment is made for permission to use the following items: the drawings on pages 25, 40, 69, and 150 are reprinted with permission by Sid Harris, drawings on pages 109, 144, and 178 were commissioned by the Committee and appear courtesy of Mike Mikula; and the drawing on page 132 is reprinted with permission from the New Yorker/Cartoon Bank.

# THE NATIONAL ACADEMIES
*Advisers to the Nation on Science, Engineering, and Medicine*

The **National Academy of Sciences** is a private, nonprofit, self-perpetuating society of distinguished scholars engaged in scientific and engineering research, dedicated to the furtherance of science and technology and to their use for the general welfare. Upon the authority of the charter granted to it by the Congress in 1863, the Academy has a mandate that requires it to advise the federal government on scientific and technical matters. Dr. Bruce M. Alberts is president of the National Academy of Sciences.

The **National Academy of Engineering** was established in 1964, under the charter of the National Academy of Sciences, as a parallel organization of outstanding engineers. It is autonomous in its administration and in the selection of its members, sharing with the National Academy of Sciences the responsibility for advising the federal government. The National Academy of Engineering also sponsors engineering programs aimed at meeting national needs, encourages education and research, and recognizes the superior achievements of engineers. Dr. Wm. A. Wulf is president of the National Academy of Engineering.

The **Institute of Medicine** was established in 1970 by the National Academy of Sciences to secure the services of eminent members of appropriate professions in the examination of policy matters pertaining to the health of the public. The Institute acts under the responsibility given to the National Academy of Sciences by its congressional charter to be an adviser to the federal government and, upon its own initiative, to identify issues of medical care, research, and education. Dr. Harvey V. Fineberg is president of the Institute of Medicine.

The **National Research Council** was organized by the National Academy of Sciences in 1916 to associate the broad community of science and technology with the Academy's purposes of furthering knowledge and advising the federal government. Functioning in accordance with general policies determined by the Academy, the Council has become the principal operating agency of both the National Academy of Sciences and the National Academy of Engineering in providing services to the government, the public, and the scientific and engineering communities. The Council is administered jointly by both Academies and the Institute of Medicine. Dr. Bruce M. Alberts and Dr. Wm. A. Wulf are chair and vice chair, respectively, of the National Research Council

**www.national-academies.org**

**Staff**

**RICHARD BISSELL,** Executive Director
**DEBORAH D. STINE,** Associate Director
**LAUREL HAAK,** Program Officer
**MARION RAMSEY,** Administrative Associate

# Preface

Over the last decade, the National Academies Committee on Science, Engineering, and Public Policy (COSEPUP) has issued a series of reports on how science and engineering are performed and supported in the United States and how future generations of scientists are trained and educated.[1] A point made by each report is that science and engineering research continually evolves beyond the boundaries of single disciplines and offers employment opportunities that require not only *depth* of knowledge but also *breadth* of knowledge, integration, synthesis, and an array of skills. Several reports suggested that a greater emphasis on interdisciplinary research and training would be consistent with those findings.

In May 2003, the National Academies and the W.M. Keck Foundation announced the National Academies Keck *Futures Initiative*, a program designed to realize the full potential of interdisciplinary research (IDR). Specifically, the *Futures Initiative* was created to "stimulate new modes of inquiry and break down the conceptual and institutional barriers to interdisciplinary research that could yield significant benefits to science and society."

As indicated by Robert A. Day, chairman and chief executive officer of the W. M. Keck Foundation, "The *Futures Initiative* is designed to create a

---

[1]See, for example, *Science, Technology, and the Federal Government: National Goals for a New Era* (1993), which emphasized the importance of human resources for the scientific enterprise, and *Reshaping the Graduate Education of Scientists and Engineers* (1995), which urged expanded training opportunities for students to prepare them not only for academic careers but also for wider employment opportunities. Later reports dealt with changing careers and mentoring students in science and engineering.

powerful, ongoing forum where the best and brightest minds from across the disciplines of science, technology, and medical research can come together and ask each other, 'What if . . . ?' More than that, they can then secure the funds necessary to pursue ideas and conduct follow-on research. Training individuals who are conversant in ideas and languages of other fields is central to the continued march of scientific progress in the 21st century. The W. M. Keck Foundation is proud to participate in this important effort."

As part of the *Futures Initiative*, the Keck Foundation asked the National Academies to review the state of interdisciplinary research and education in science and engineering and recommend ways to facilitate them. Accordingly, COSEPUP, under the aegis of the National Academies, created the Committee on Facilitating Interdisciplinary Research, whose members were drawn from government, academe, and industry and had long experience in leading and performing IDR.[2] The committee was charged with the following tasks:

1. Review proposed definitions of interdisciplinary research, including similarities and differences from research characterized as cross-disciplinary, intradisciplinary, and multidisciplinary, and develop measures to determine whether research is interdisciplinary or not.

2. Identify and analyze current structural models of interdisciplinary research.

3. Identify and analyze the policies and procedures of Congress, funding organizations, and institutions that encourage or discourage interdisciplinary research.

4. Compare and contrast current structural models and policies and procedures in academic and nonacademic settings as well as traditional and nontraditional academic settings that encourage or discourage interdisciplinary research.

5. Identify measures that can be used to evaluate the impact on research, graduate students and postdoctoral scholars, and researchers expected from their engagement in greater interdisciplinary research and cross-professional opportunities.

6. Develop findings and conclusions as to the current state of interdisciplinary research and the factors that encourage (or discourage) it in academic, industry, and federal laboratory settings.

7. Provide recommendations to academic institutions and public and private sponsors of research as to how to better stimulate and support interdisciplinary research.

---

[2]Biographical information on members of the committee are listed in Appendix A.

The committee's methods and the framework for this report are provided in the "Note to the Reader" that follows the Executive Summary. In sum, the committee based its analysis of how to facilitate IDR on its Convocation on Facilitating IDR, surveys, focus groups, interviews with scholars, and an extensive literature review.

The committee was hampered in its attempt to compare models and policies that encourage IDR by a lack of recent published information. There is a considerable history of research, but the committee found insufficient evidence to answer such questions as, Which, if any, emerging IDR fields and subfields should be strengthened? What technologies and instruments are most likely to generate new ID fields and subfields? Where (if anywhere) should the government increase its investment in IDR? This report is the latest in a growing literature on models and policies that situates the discussion in the current context of science and engineering, and it formally recommends increased research to provide the necessary answers.

Similarly, in attempting to compare academic and nonacademic research practices, the committee found substantial asymmetries. Interdisciplinarity has long been accepted and familiar in many industrial and government laboratories and other nonacademic settings; such settings traditionally emphasize teams and problem-driven research, and they permit researchers to move easily between laboratories, to share their skills, and to acquire new ones. In academe, however, such collaboration is often impeded by administrative, funding, and cultural barriers between departments, by which most research and teaching activities are organized. For that reason and because the highest concentration of scholarly expertise is found in universities, this report focuses primarily on facilitating IDR in academe.

The study identified academic institutional customs that create a small but persistent "drag" on researchers who would like to do interdisciplinary research and teaching. They include especially the academic promotion and reward system and the department-based budgeting structures of universities. The committee concluded that IDR nevertheless plays an essential and growing role in permitting researchers to venture beyond the frontiers of their own disciplines and address questions of ever-increasing complexity and societal urgency. The committee identified "best practices" identified in its investigation that can be applied by those who wish to facilitate IDR, including undergraduate and graduate students, postdoctoral fellows, faculty members, researchers, funding organizations, academic and nonacademic institutions, and disciplinary societies. In some of the cases, institutions have experimented with substantial alteration of the traditional academic structures or even replacement with new structures and models to reduce barriers to IDR. It also found that improved evaluation tools, such as the ability to provide a broader peer review of interdisciplinary proposals and publication submissions, can greatly assist those who wish to conceptualize, fund, and administer IDR. More best practices, of course, exist than are provided in this report.

In conclusion, this report is a "call to action" for all those who perform, administer, support, and organize interdisciplinary research and training. Its purpose is to facilitate collaborative practices that can increase the productivity of science and engineering. The majority of the report suggests "incremental" changes that will facilitate interdisciplinary research. In Chapter 9, however, the committee provides suggestions for "transformative" changes for those institutions who are willing to experiment with new approaches. Research partnerships must be especially tailored to address scientific and societal challenges in innovative ways. The purpose of this report is not to privilege the pursuit of IDR over disciplinary research, but rather to seek to provide suggestions to those interested or engaged in interdisciplinarity to optimize its effectiveness and strengthen both IDR and the disciplinary foundations from which it springs.

Nancy C. Andreasen
Theodore L. Brown
*Co-Chairs*
Committee on Facilitating Interdisciplinary Research

# Acknowledgments

This report is the product of many people. First, we thank all those who spoke at our convocation in January 2004. They were (in alphabetical order)

**ANTHONY ARMSTRONG**, Director, Indiana 21st Century Research & Technology Fund

**RUZENA BAJCSY**, Director, Center for Information Technology Research in the Interest of Society, University of California, Berkeley

**WILLIAM BERRY**, Director, Basic Research, ODUSD, Department of Defense

**MARYE ANNE CARROLL**, Professor, Atmospheric, Oceanic, and Space Sciences; Professor, Chemistry; Director, Program for Research on Oxidants: Photochemistry, Emissions, and Transport (PROPHET); Director, Biosphere-Atmosphere Research and Training (BART); University of Michigan

**CARMEN CHARETTE**, Senior Vice President, Canada Foundation for Innovation

**UMA CHOWDHRY**, Vice President, Central Research and Development, DuPont

**HARVEY COHEN**, Professor, Pediatrics, Stanford School of Medicine, and Chair, The Interdisciplinary Initiatives Committee, Bio-X, Stanford University

**JOEL E. COHEN**, Abby Rockefeller Mauzé Professor, Laboratory of Populations, Rockefeller University and Columbia University

JAMES P. COLLINS, Virginia M. Ullman Professor of Natural History and the Environment, Arizona State University

RITA R. COLWELL, Director, National Science Foundation

CLIFFORD GABRIEL, Deputy Associate Director, Science Division, White House Office of Science and Technology Policy

LAURIE R. GARDUQUE, Program Director for Research, John D. and Catherine T. MacArthur Foundation

BARRY GOLD, Program Officer, Conservation and Science, The David and Lucile Packard Foundation

ALICE GOTTLIEB, Professor of Medicine and Director of the Clinical Research Center, UMDNJ Robert Wood Johnson Medical School, University of Medicine and Dentistry New Jersey

ROBERT GRANGER, President, William T. Grant Foundation

VICTORIA INTERRANTE, Assistant Professor, Computer Science and Engineering, University of Minnesota

JULIE THOMPSON KLEIN, Professor of Humanities, Wayne State University

LINDA J. (LEE) MAGID, Professor, Chemistry, University of Tennessee, and Acting Director, Joint Institute for Neutron Sciences, UT and Oak Ridge National Laboratory

EDWARD L. MILES, Professor of Marine Studies and Public Affairs, University of Washington

MARVIN SINGER on behalf of RAY L. ORBACH, Director, Office of Science, Department of Energy

JULIO DE PAULA, Professor of Chemistry, Haverford College

MARIA PELLEGRINI, Program Director for Science, Engineering, and Liberal Arts, W. M. Keck Foundation

FENIOSKY PEÑA-MORA, Associate Professor of Construction Management and Information Technology, William E. O'Neil Faculty Scholar, Civil and Environmental Engineering Department, University of Illinois, Urbana-Champaign

DIANA RHOTEN, Program Officer, Social Science Research Council

CATHERINE ROSS, Director, Center for Quality Growth, Georgia Institute of Technology

F. SHERWOOD ROWLAND, Bren Research Professor, Chemistry and Earth System Science, University of California at Irvine

LAWRENCE A. TABAK, Director, National Institutes of Dental and Craniofacial Research, National Institute of Health

JEFFREY WADSWORTH, Director, Oak Ridge National Laboratory

PIERRE WILZIUS, Director, Beckman Institute for Advanced Science and Technology, and Professor, Materials Science and Engineering Department and Physics Department, University of Illinois at Urbana-Champaign

Without the input of each of these speakers, this report would not have been possible.

Next, we would like to thank the reviewers of this report. This report has been reviewed in draft form by individuals chosen for their diverse perspectives and technical expertise, in accordance with procedures approved by the NRC's Report Review Committee. The purpose of this independent review is to provide candid and critical comments that will assist the institution in making its published report as sound as possible and to ensure that the report meets institutional standards for objectivity, evidence, and responsiveness to the study charge. The review comments and draft manuscript remain confidential to protect the integrity of the deliberative process.

We wish to thank the following individuals for their review of this report: John Armstrong, IBM (Retired); William Brinkman, Princeton University; Norman Burkhard, Lawrence Livermore National Laboratory; Carmen Charette, Canada Foundation for Innovation; James Collins, Arizona State University; Rita Colwell, National Science Foundation; Marilyn Fogel, Carnegie Institute; Robert Frosch, Harvard University; Hedvig Hricak, Sloan-Kettering Cancer Center; Victoria Interrante, University of Minnesota; Leah Jamieson, Purdue University; Edward L. Miles, University of Washington; Diana Rhoten, Social Science Research Council; Douglas Richardson, Association of American Geographers; Dean Keith Simonton, University of California, Davis; Richard Stein, University of Massachusetts; Julie Thompson Klein, Wayne State University; Jeffrey Wadsworth, Oak Ridge Laboratory; George E. Walker, Indiana University.

Although the reviewers listed above have provided many constructive comments and suggestions, they were not asked to endorse the conclusions or recommendations, nor did they see the final draft of the report before its release. The review of this report was overseen by Rebecca Chopp, Colgate University, and Pierre Hohenberg, New York University. Appointed by the National Research Council, they were responsible for making certain that an independent examination of this report was carried out in accordance with institutional procedures and that all review comments were carefully considered. Responsibility for the final content of this report rests entirely with the authoring committee and the institution.

In addition, we would like to thank Maxine Singer, the chair of COSEPUP and the guidance group that oversaw this project which included:

**JAMES DUDERSTADT** (Guidance Group Chair), President Emeritus and University Professor of Science and Engineering, University of Michigan

**MARY-CLARE KING,** American Cancer Society Professor of Medicine
and Genetics, University of Washington
**GERALD M. RUBIN,** Vice President for Biomedical Research, Howard
Hughes Medical Institute
**EDWARD H. SHORTLIFFE,** Professor and Chair, Department of
Medical Informatics, Columbia University, Vanderbilt Clinic,
Columbia Presbyterian Medical Center
**MAXINE SINGER,** President Emeritus, Carnegie Institution of
Washington

Finally, we would like to thank the staff for this project, including
Deborah Stine, associate director for the Committee on Science, Engineering, and Public Policy and study director, who managed the project; Laurel
Haak, program officer with the Committee on Science, Engineering, and
Public Policy who conducted interviews, wrote boxes, organized the convocation, and conducted research and analysis; Alan Anderson, the science
writer for this report; Erin McCarville and Camille Collett, who provided
project support; Christine Mirzayan Science and Technology Policy Fellows
Heather Agler, Mary Anderson, Mary Feeney, Jesse Gray, Rebecca Janes,
Joshua Schnell, and Gretchen Schwarz, who all provided research and
analytical support; and Richard Bissell, executive director of the Committee
on Science, Engineering, and Public Policy and of Policy and Global Affairs.

# Contents

## 10 FINDINGS AND RECOMMENDATIONS                                    188

## APPENDIXES

# Figures, Tables, and Boxes

## FIGURES

# TABLES

# BOXES

## Innovative Practice

## Toolkit

## Definition

## Evolution

### Structures/Policies

# Executive Summary

Interdisciplinary research (IDR) can be one of the most productive and inspiring of human pursuits—one that provides a format for conversations and connections that lead to new knowledge. As a mode of discovery and education, it has delivered much already and promises more—a sustainable environment, healthier and more prosperous lives, new discoveries and technologies to inspire young minds, and a deeper understanding of our place in space and time. Despite the apparent benefits of IDR, researchers interested in pursuing it often face daunting obstacles and disincentives. Some of them take the form of personal communication or "culture" barriers; others are related to the tradition in academic institutions of organizing research and teaching activities by discipline-based departments—a tradition that is commonly mirrored in funding organizations, professional societies, and journals.

Under the sponsorship of the Keck Foundation, the National Academies Committee on Facilitating Interdisciplinary Research examined the scope of IDR. It drew conclusions and made recommendations based on the committee's deliberations and on suggestions it received from undergraduate and graduate students, postdoctoral scholars, researchers, academic and nonacademic institutional leaders, funding organizations, and professional societies at its convocation and via its survey; the focus groups held at the National Academies Keck *Futures Initiative* Conference; and interviews with leading scholars.

The recommendations proposed here can help students, postdoctoral scholars, researchers, institutions, funding organizations, professional societies, and those who evaluate research to help IDR to reach its full potential.

## FINDINGS

The committee's 15 findings are organized here in three categories: the definition of IDR, its current situation, and the changes needed to facilitate it.

### Definition

1.  Interdisciplinary research (IDR) is a mode of research by teams or individuals that integrates information, data, techniques, tools, perspectives, concepts, and/or theories from two or more disciplines or bodies of specialized knowledge to advance fundamental understanding or to solve problems whose solutions are beyond the scope of a single discipline or area of research practice.

### Current Situation

2.  IDR is pluralistic in method and focus. It may be conducted by individuals or groups and may be driven by scientific curiosity or practical needs.

3.  Interdisciplinary thinking is rapidly becoming an integral feature of research as a result of four powerful "drivers": the inherent complexity of nature and society, the desire to explore problems and questions that are not confined to a single discipline, the need to solve societal problems, and the power of new technologies.

4.  Successful interdisciplinary researchers have found ways to integrate and synthesize disciplinary depth with breadth of interests, visions, and skills.

5.  Students, especially undergraduates, are strongly attracted to interdisciplinary courses, especially those of societal relevance.

6.  The success of IDR groups depends on institutional commitment and research leadership. Leaders with clear vision and effective communication and team-building skills can catalyze the integration of disciplines.

### Challenges to Overcome

7.  The characteristics of IDR pose special challenges for funding organizations that wish to support it. IDR is typically collaborative and

involves people of disparate backgrounds. Thus, it may take extra time for building consensus and for learning new methods, languages, and cultures.

8. Social-science research has not yet fully elucidated the complex social and intellectual processes that make for successful IDR. A deeper understanding of these processes will further enhance the prospects for creation and management of successful IDR programs.

## Changes Needed

9. In attempting to balance the strengthening of disciplines and the pursuit of interdisciplinary research, education, and training, many institutions are impeded by traditions and policies that govern hiring, promotion, tenure, and resource allocation.

10. The increasing specialization and cross-fertilizations in science and engineering require new modes of organization and a modified reward structure to facilitate interdisciplinary interactions.

11. Professional societies have the opportunity to facilitate IDR by producing state-of-the-art reports on recent research developments and on curriculum, assessment, and accreditation methods; enhancing personal interactions; building partnerships among societies; publishing interdisciplinary journals and special editions of disciplinary journals; and promoting mutual understanding of disciplinary methods, languages, and cultures.

12. Reliable methods for prospective and retrospective evaluation of interdisciplinary research and education programs will require modification of the peer-review process to include researchers with interdisciplinary expertise in addition to researchers with expertise in the relevant disciplines.

## Lessons from Industry and National Laboratories

13. Industrial and national laboratories have long experience in supporting IDR. Unlike universities, industry and national laboratories organize by the problems they wish their research enterprise to address. As problems come and go, so does the design of the organization.

14. Although research management in industrial and government settings tends to be more "top-down" than it is at universities, some of its lessons may be profitably incorporated into universities' IDR strategies.

15. Collaborative interdisciplinary research partnerships among universities, industry, and government have increased and diversified rapidly. Although such partnerships still face significant barriers, well-documented studies provide strong evidence of both their research benefits and their effectiveness in bringing together diverse cultures.

## RECOMMENDATIONS

On the basis of its findings, the committee offers the following recommendations. They are listed by category of people and organizations involved in interdisciplinary research, education, and training. The committee does not necessarily urge interdisciplinary research activities for all institutions and individuals, but, for parties that are interested in implementing or improving such activities, the committee provides the following recommendations.

The majority of the recommendations the committee makes to facilitate interdisciplinary research are "incremental"; however, the committee provides suggestions for "transformative" changes for those institutions willing to experiment with new approaches. Most of these are described briefly here in the section entitled "academic institutional structures," but very specific ideas are provided in Chapter 9 that expand upon these recommendations.

### Students

S-1: *Undergraduate students* should seek out interdisciplinary experiences, such as courses at the interfaces of traditional disciplines that address basic research problems, interdisciplinary courses that address societal problems, and research experiences that span more than one traditional discipline.

S-2: *Graduate students* should explore ways to broaden their experience by gaining "requisite" knowledge in one or more fields in addition to their primary field.

### Postdoctoral Scholars

P-1: Postdoctoral scholars can actively exploit formal and informal means of gaining interdisciplinary experiences during their postdoctoral appointments through such mechanisms as networking events and internships in industrial and nonacademic settings.

P-2: Postdoctoral scholars interested in interdisciplinary work should seek to identify institutions and mentors favorable to IDR.

### Researchers and Faculty Members

R-1: Researchers and faculty members desiring to work on interdisciplinary research, education, and training projects should immerse themselves in the languages, cultures, and knowledge of their collaborators in IDR.

R-2: Researchers and faculty members who hire postdoctoral scholars from other fields should assume the responsibility for educating them in the new specialties and become acquainted with the postdoctoral scholars' knowledge and techniques.

## Educators

A-1: Educators should facilitate IDR by providing educational and training opportunities for undergraduates, graduate students, and postdoctoral scholars, such as relating foundation courses, data gathering and analysis, and research activities to other fields of study and to society at large.

## Academic Institutions' Policies

I-1: Academic institutions should develop new and strengthen existing policies and practices that lower or remove barriers to interdisciplinary research and scholarship, including developing joint programs with industry and government and nongovernment organizations.

I-2: Beyond the measures suggested in I-1, institutions should experiment with more innovative policies and structures to facilitate IDR, making appropriate use of lessons learned from the performance of IDR in industrial and national laboratories.

I-3: Institutions should support interdisciplinary education and training for students, postdoctoral scholars, researchers, and faculty by providing such mechanisms as undergraduate research opportunities, faculty team-teaching credit, and IDR management training.

I-4: Institutions should develop equitable and flexible budgetary and cost-sharing policies that support IDR.

## Team Leaders

T-1: To facilitate the work of an IDR team, its leaders should bring together potential research collaborators early in the process and work toward agreement on key issues.

T-2: IDR leaders should seek to ensure that each participant strikes an appropriate balance between leading and following and between contributing to and benefiting from the efforts of the team.

Funding Organizations

F-1: Funding organizations should recognize and take into consideration in their programs and processes the unique challenges faced by IDR with respect to risk, organizational mode, and time.

F-2: Funding organizations, including interagency cooperative activities, should provide mechanisms that link interdisciplinary research and education and should provide opportunities for broadening training for researchers and faculty members.

F-3: Funding organizations should regularly evaluate, and if necessary redesign, their proposal and review criteria to make them appropriate for interdisciplinary activities.

F-4: Congress should continue to encourage federal research agencies to be sensitive to maintaining a proper balance between the goal of stimulating interdisciplinary research and the need to maintain robust disciplinary research.

Professional Societies

PS-1: Professional societies should seek opportunities to facilitate IDR at regular society meetings and through their publications and special initiatives.

Journal Editors

J-1: Journal editors should actively encourage the publication of IDR research results through various mechanisms, such as editorial-board membership and establishment of special IDR issues or sections.

Evaluation of IDR

E-1: IDR programs and projects should be evaluated in such a way that there is an appropriate balance between criteria characteristic of IDR, such as contributions to creation of an emerging field and whether they lead to practical answers to societal questions, and traditional disciplinary criteria, such as research excellence.

E-2: Interdisciplinary education and training programs should be evaluated according to criteria specifically relevant to interdisciplinary activities, such as number and mix of general student population participation and knowledge acquisition, in addition to the usual requirements of excellence in content and presentation.

E-3: Funding organizations should enhance their proposal-review mechanisms so as to ensure appropriate breadth and depth of expertise in the review of proposals for interdisciplinary research, education, and training activities.

E-4: Comparative evaluations of research institutions, such as the National Academies' assessment of doctoral programs and activities that rank university departments, should include the contributions of interdisciplinary activities that involve more than one department (even if it involves double-counting), as well as single-department contributions.

## Academic Institutional Structure

U-1: Institutions should explore alternative administrative structures and business models that facilitate IDR across traditional organizational structures.

U-2: Allocations of resources from high-level administration to interdisciplinary units, to further their formation and continued operation, should be considered in addition to resource allocations of discipline-driven departments and colleges. Such allocations should be driven by the inherent intellectual values of the research and by the promise of IDR in addressing urgent societal problems.

U-3: Recruitment practices, from recruitment of graduate students to hiring of faculty members, should be revised to include recruitment across department and college lines.

U-4: The traditional practices and norms in hiring of faculty members and in making tenure decisions should be revised to take into account more fully the values inherent in IDR activities.

U-5: Continuing social science, humanities, and information-science-based studies of the complex social and intellectual processes that make for successful IDR are needed to deepen the understanding of these processes and to enhance the prospects for the creation and management of successful programs in specific fields and local institutions.

## A NOTE TO THE READER

This report addresses five primary populations, all of whom participate in interdisciplinary research (IDR): researchers and educators, undergraduate and graduate students and postdoctoral scholars, institutions, private and federal organizations that fund research and education, and professional societies.

At the risk of some repetition, the guide addresses the primary groups in separate sections because of differences in perspective, primary objectives, and responsibilities.

### Organization of the Report

Prominent in the discussion in this report is an analysis of what facilitates—and what impedes—interdisciplinary research. The report is organized as follows:

- **Chapter 1** provides an "interdisciplinary vision" and describes where the research community has been and where it is going.
- **Chapter 2** provides a definition of IDR, discusses four driving forces of IDR, and explores the nature of successful interdisciplinary work.
- **Chapter 3** provides several case studies describing how interdisciplinary research is performed in industry and national laboratories. Although the major emphasis in this study is on the state of IDR in academic institutions, IDR plays important roles in industrial and government laboratories, and an understanding of the drivers for IDR in those settings can provide helpful insights in the examination of IDR in academic settings.
- **Chapter 4** describes the current working environment and challenges for individual students and academic researchers interested in IDR.
- **Chapter 5** discusses the institutional barriers to interdisciplinary education and research and discusses possible research, education, and training policies to facilitate interdisciplinary work.
- **Chapter 6** discusses the barriers that federal and private funding organizations encounter in their support of interdisciplinary education and research activities and proposes some innovative funding strategies.
- **Chapter 7** discusses the role that professional societies play in facilitating interdisciplinary education and research.
- **Chapter 8** describes the challenges of evaluating interdisciplinary research and education activities, including evaluating the direct and indirect impacts of IDR; the people who perform IDR; the institutions, centers, and programs that engage in IDR; and the issue of national comparative assessment of departments.
- **Chapter 9** examines the overall structures in which IDR takes place and proposes some incremental and transformative policies to facilitate it.

- **Chapter 10** synthesizes the committee's findings and recommendations (also presented at the end of each chapter) to provide an overarching picture of the actions that can be taken by all the populations described to facilitate interdisciplinary research and education.

## Method

The work of the committee began with a review of the literature—the results of which are provided in Appendix H.

The committee also undertook a number of activities to collect additional information; these are described in several appendixes:

- **Appendix C** provides additional information on the Convocation on Facilitating Interdisciplinary Research hosted by the committee on January 29-30, 2004 in Washington, D.C. At the convocation, the committee heard the experiences and opinions of representatives from private, federal, international, and state funding organizations who have had leading roles in facilitating IDR; leading senior and junior researchers involved in IDR; interdisciplinary research-center directors; experts in interdisciplinary education and training; and more than 200 participants.

In addition, the convocation included a poster session that featured some 30 model interdisciplinary programs and opportunities for participants to provide their thoughts to the committee in written (survey) and oral form.

References to speaker presentations and convocation participant comments appear throughout the report.

- **Appendix D** provides a qualitative and quantitative historical analysis of the development of IDR and interdisciplines, university departments, and professional societies.
- **Appendix E** provides an analysis of the committee's surveys of students, postdoctoral scholars, faculty members, funders, policy makers, and disciplinary societies involved in interdisciplinary research and education. This analysis is referred to throughout the report. The surveys asked questions about the impediments, programs, and evaluation criteria related to IDR and gathered suggestions for recommendations on how to facilitate IDR.

The first survey, referred to in the report as the "convocation survey," was given to participants who attended the convocation described above; 91 convocation participants responded to the survey.

A slightly modified version of the convocation survey, called the "individual survey," was posted on the committee Web site. An invitation to participate in the survey was sent to universities, professional societies, nongovernment organizations, and participants in federal and private interdisciplinary programs; 423 people responded to the solicitation.

An invitation to participate in a third survey, called the "provost survey," was distributed on line to provosts or vice-chancellors of institutions that conduct IDR; 57 institutions responded.

• **Appendix F** provides a list of the administrators, scholars, and center directors interviewed by the committee and summarizes the thoughts they offered regarding IDR.

• **Appendix G** summarizes the statements of interdisciplinary researchers in a wide variety of research fields who participated in three focus groups at the first Keck *Futures* Conference, titled "Signals, Decisions, and Meaning in Biology, Chemistry, Physics, and Engineering," held on November 14 in Irvine, California.

• **Appendix H** provides the report bibliography.

## Boxes

Throughout this report, text boxes are used to highlight activities, programs, and policies that the committee found particularly interesting and to help to illustrate its findings and recommendations. These boxes are summaries of existing literature and reports or are based on new information gathered by the committee. They are organized into seven categories:

• **Innovative Practices** highlight existing programs that are particularly innovative and that illustrate the committee's recommendations.

• **Structures and Policies** illustrate unique organizational structures and institutional policies.

• **Toolkit** provides illustrations of how proposals, individuals, funding organization programs, interdisciplinary centers, and research outcomes can be evaluated.

• **Definitions** describe and define IDR, its management, and its evaluation.

• **Evolution** shows how research, organizations, and institutions involved in IDR have changed.

• **Convocation Quotes** are snapshots of particularly revealing or insightful comments by panelists and participants of the convocation that illustrate some of the key barriers and drivers for IDR.

• **Survey Analysis** provides quantitative highlights from the committee's surveys of convocation participants and others.

## Case Table

To help the reader navigate the case studies presented in the report, Table ES-1 provides a table of the boxes in the report, listed in order of appearance, by category and title. For each box, the major topics are indicated. Most boxes cover more than one topic area.

- **Driver:** These boxes illustrate the four drivers of IDR, the inherent complexity of nature (C), the drive to explore basic research at the interfaces (I), the need to solve societal problems (S), and the stimulus of generative technologies (G).
- **Industry:** These boxes show how industry plays a role in IDR.
- **National Lab:** These boxes provide examples of IDR at national labs.
- **Academe:** In these boxes, IDR in academic settings is highlighted.
- **Undergrad, Graduate, Postdoc, and Faculty:** These boxes provide examples of programs and policies to facilitate interdisciplinary work for these groups of students, researchers, and teachers.
- **Structure:** These boxes show how particular administrative and bricks and mortar structures can facilitate IDR.
- **Policy:** These boxes provide discrete examples of effective policies to promote interdisciplinary work.
- **Evaluation:** These boxes illustrate a variety of strategies for evaluating interdisciplinary people and programs.
- **Funding:** These boxes show how funding agencies have effectively facilitated IDR.
- **History:** These boxes provide a historical overview of particular interdisciplinary projects or fields.
- **Managing Collaborations:** These boxes illustrate management options for bringing together and maintaining interdisciplinary teams.
- **Professional Society:** These boxes show how professional societies have played a role fostering and facilitating IDR.

The committee hopes that this report will increase the understanding of interdisciplinary research and encourages readers to undertake actions that will help facilitate it.

**TABLE ES-1** List of Boxes by Order of Appearance, by Category and Title

| Box | Category | Case/Topic |
|-----|----------|-----------|
| 1-1 | Struct/Policy | Columbia Univ./ Brown Univ. |
| 1-2 | Struct/Policy | IDR in Netherlands |
| 1-3 | Struct/Policy | EURAB Report |
| 2-2 | Evolution | MIT Radiation Laboratory |
| 2-3 | Evolution | X-Ray Crystallography |
| 2-4 | Innovative Practice | KDI Institute |
| 2-5 | Evolution | Argonne Nat'l Labs Advanced Photon Source |
| 3-1 | Innovative Practice | Philips Physics Research Laboratory |
| 3-2 | Innovative Practice | Role of IDR at IBM |
| 3-3 | Innovative Practice | Hard-Disk-Drive Research |
| 4-1 | Toolkit | Summer Research Opportunities |
| 4-2 | Innovative Practice | Arizona State Univ. School of Life Sciences |
| 4-3 | Innovative Practice | Harvard Univ. Global Assessment Project |
| 4-4 | Innovative Practice | Univ. Minnesota, Institute for Mathematics and its Applications |
| 4-5 | Innovative Practice | Penn State University, Huck Institutes |
| 4-6 | Innovative Practice | Fred Hutchinson Cancer Research Center |
| 5-1 | Evolution | NRC Graduate Program Assessment |
| 5-2 | Innovative Practice | Physical Barriers to IDR |
| 5-3 | Innovative Practice | Haverford College |
| 5-4 | Innovative Practice | University of Wisconsin |
| 5-5 | Toolkit | University of Southern California |
| 5-6 | Toolkit | Univ. Illinois Urbana-Champaign, Beckman Institute |
| 5-7 | Toolkit | State University of NY, Stony Brook |
| 5-8 | Toolkit | UC Davis, Univ. Michigan |
| 6-1 | Evolution | DARPA |
| 6-2 | Innovative Practice | NASA — NAI |

| Driver | Industry | National Lab | Academe | Undergrad | Graduate | Postdoc | Faculty | Structure | Policy | Evaluation | Funding | History | Managing Collaborations | Prof. Society |
|---|---|---|---|---|---|---|---|---|---|---|---|---|---|---|
| | | | X | | | | | X | | | | | | |
| | | | X | X | X | | | X | X | X | | | X | |
| | | | X | X | X | | | X | X | X | | | X | |
| S | X | X | X | | | | | | | | | X | | |
| G | | | | | | | | | | | | X | | |
| G | | | X | | | | | | | X | X | | | X | |
| G | X | X | X | | | | | | | | | | X | |
| | X | | | | | | | | | | | | X | |
| | X | | | | | | | | | | | | X | |
| | X | | X | | | | | | | | | | X | |
| | X | X | X | X | X | X | X | | | | | | | |
| | | | X | X | X | | X | | | | | | X | |
| | | | X | | | X | | | | X | | | X | |
| | X | | X | | | X | | | | X | X | | | |
| | | | X | X | X | | X | X | | X | | | X | |
| | | | X | | X | X | X | | | | | | X | |
| | | | X | | X | | | | | X | | | | |
| | X | X | X | | | | | X | | | | | X | |
| | | | X | X | | | | | | | | | | |
| | | | X | | | | X | | X | X | | | | |
| | | | X | | | | X | | X | X | | | | |
| | | | X | | | | X | | X | X | X | | | |
| | | | X | | | | X | | | | | | X | |
| | | | X | | | | | | X | | X | | | |
| S | | | | | | | | | | | X | X | X | |
| I | | | X | | X | X | X | | | | X | | X | |

*continues*

**TABLE ES-1** Continued

| Box | Category | Case/Topic |
|-----|----------|-----------|
| 6-3 | Innovative Practice | NIH |
| 6-4 | Innovative Practice | DoD — MURI |
| 6-5 | Innovative Practice | BWF — Career Transition Awards |
| 6-6 | Evolution | Rice University |
| 6-7 | Innovative Practice | HHMI — Janelia Farm |
| 6-8 | Toolkit | OSTP |
| 6-9 | Evolution | Biomedical Engineering |
| 7-1 | Toolkit | Journals |
| 7-2 | Toolkit | Professional Societies |
| 7-3 | Innovative Practice | Assn. of American Geographers |
| 7-4 | Innovative Practice | Coalition for Bridging the Sciences |
| 8-1 | Toolkit | Harvard Interdisciplinary Studies Project |
| 8-2 | Innovative Practice | National Science Foundation Engineering Research Centers |
| 8-3 | Evolution | Hybrid Vigor Institute |
| 8-4 | Toolkit | National Science Foundation IGERT |
| 8-5 | Toolkit | Dutch Universities |
| 8-6 | Toolkit | Transdisciplinary Tobacco Use Research Centers |
| 9-1 | Definition | Matrix Management |
| 9-2 | Innovative Practice | Evergreen State College, Penn State Univ., Harvard Univ., Brown Univ. |
| 9-3 | Innovative Practice | Rockefeller University |
| 9-4 | Innovative Practice | Purdue University |
| 9-5 | Innovative Practice | Univ. Washington Program on the Environment, CMU/University Pittsburgh Center for Neural Basis of Cognition |
| 9-6 | Innovative Practice | Stanford University Bio-X |
| 9-7 | Innovative Practice | Biomedical Informatics Research Network |

| Driver | Industry | National Lab | Academe | Undergrad | Graduate | Postdoc | Faculty | Structure | Policy | Evaluation | Funding | History | Managing Collaborations | Prof. Society |
|---|---|---|---|---|---|---|---|---|---|---|---|---|---|---|
| I |  |  |  |  | X | X | X |  |  |  | X |  | X |  |
| S |  |  | X |  | X | X | X |  |  |  | X |  | X |  |
| I |  |  |  |  |  | X | X |  |  |  | X |  | X |  |
| G |  |  | X |  |  |  |  | X |  |  |  | X |  |  |
| I |  |  | X |  |  |  |  | X |  |  | X |  | X |  |
|  |  |  |  |  |  |  |  | X | X |  | X |  |  |  |
| I |  |  | X |  |  |  |  |  |  | X | X |  | X | X |
|  |  |  |  |  |  |  |  |  |  |  |  |  | X | X |
|  |  |  |  |  |  | X | X |  |  | X | X |  | X | X |
|  | X |  | X |  |  |  |  | X | X |  | X | X | X | X |
|  |  |  |  |  |  |  |  |  |  |  | X |  | X | X |
|  |  |  | X |  |  |  | X |  |  | X |  |  |  |  |
| I | X |  | X |  |  |  |  | X | X | X | X |  | X |  |
|  |  |  | X |  | X | X | X | X |  |  | X |  | X |  |
| I |  |  | X |  | X | X |  | X |  |  | X | X | X |  |
|  |  |  | X |  |  |  |  |  |  | X |  |  |  |  |
|  |  |  | X |  |  |  |  |  |  | X |  |  | X |  |
|  |  |  | X |  |  |  |  | X | X |  |  |  | X |  |
|  |  |  | X | X |  |  |  |  |  | X |  |  |  |  |
|  |  |  | X |  |  |  |  | X | X |  |  | X | X |  |
| I |  |  | X |  |  |  |  | X | X | X |  |  |  |  |
|  |  |  | X |  |  |  |  | X | X | X |  |  | X |  |
| I |  |  | X |  |  |  |  | X | X |  |  |  | X |  |
| G |  |  | X |  |  |  |  | X |  | X |  |  | X |  |

# 1

# A Vision of
# Interdisciplinary Research

Interdisciplinary research[1] (IDR) can be one of the most productive and inspiring of human pursuits—one that provides a format for conversations and connections that lead to new knowledge. As a mode of discovery and education, it has delivered much already and promises more—a sustainable environment, healthier and more prosperous lives, new discoveries and technologies to inspire young minds, and a deeper understanding of our place in space and time.

> We are not students of some subject matter, but students of problems. And problems may cut right across the borders of any subject matter or discipline.
>
> Karl Popper[2]

Interdisciplinary research and education are inspired by the drive to solve complex questions and problems, whether generated by scientific curiosity or by society, and lead researchers in different disciplines to meet at the interfaces and frontiers of those disciplines and even to cross frontiers to form new disciplines.

---

[1]The definition of IDR is provided and discussed in Chapter 2.
[2]Popper, K. R. Conjectures and Refutations: The Growth of Scientific Knowledge. New York: Routledge and Kegan Paul, 1963, p. 88.

The history of science from the time of the earliest scholarship abounds with examples of the integration of knowledge from many research fields. The pre-Socratic philosopher Anaximander brought together his knowledge of geology, paleontology, and biology to discern that living beings develop from simpler to more complex forms. In the age of the great scientific revolutions of 17th-century Europe, its towering geniuses—Isaac Newton, Robert Hooke, Edmond Halley, Robert Boyle, and others—brought their curiosity to bear not only on subjects that would lead to basic discoveries that bear their names but also on every kind of interdisciplinary challenge, including military and mining questions.[3] In the 19th century, Louis Pasteur became a model interdisciplinarian, responding to practical questions about diseases and wine spoilage with surprising answers that laid the foundations of microbiology and immunology. Today, the proliferation of new understanding about the molecular and genetic underpinnings of life demonstrates the power of combining disciplinary knowledge in interdisciplinary ways.

In recent decades, the growth of scientific and technical knowledge has prompted scientists, engineers, social scientists, and humanists to join in addressing complex problems that must be attacked simultaneously with deep knowledge from different perspectives. Students show increasing enthusiasm about problems of global importance that have practical consequences, such as disease prevention, economic development, social inequality, and global climate change—all of which can best be addressed through IDR. A glance across the research landscape reveals how many of today's "hot topics" are interdisciplinary: nanotechnology, genomics and proteomics,[4] bioinformatics, neuroscience, conflict, and terrorism. All those invite and even demand interdisciplinary participation. Similarly, many of the great research triumphs are products of interdisciplinary inquiry and collaboration: discovery of the structure of DNA, magnetic resonance imaging, the Manhattan Project, laser eye surgery, radar, human genome sequencing, the "green revolution," and manned space flight. There can be no question about the productivity and effectiveness of research teams formed of partners with diverse expertise.[5]

---

[3]Robert K. Merton's classic *Science, Technology and Society in Seventeenth Century England* describes the work of the remarkable "natural philosophers" whose reach spanned many of today's disciplines.

[4]Study of all the proteins encoded by an organism's DNA.

[5]A recent editorial in *Science* notes, "The time is upon us to recognize that the new frontier is the interface, wherever it remains unexplored. . . . In the years to come, innovators will need to jettison the security of familiar tools, ideas, and specialties as they forge new partnerships." Kafatos, F.C. and Eisner, T., "Unification in the Century of Biology." *Science*, 303 (February 27):1257, 2004.

On an individual basis, studies[6] show that situational factors, such as exposure to ideas outside one's own discipline, may have a positive impact on researchers in their own discipline. Prolific and influential researchers are more likely to keep up with developments outside their own domains, and this interdisciplinary curiosity can lead to major breakthroughs on their own projects. For example, it was Charles Darwin's reading of Malthus's "An Essay on the Principle of Population" that led to his theory of natural selection.

---

**Convocation Quote**

One of the things that I have observed is how increasingly the fields of sociology, bioethics, and economics are necessary to execute our missions in the apparently harder sciences as we move ahead.

Jeffrey Wadsworth, director, Oak Ridge National Laboratory

---

Academe has responded to the burgeoning specialization of knowledge and increased cross-fertilization by creating new hybrid research fields— such as bioengineering, biogeochemistry, and paleoseismology—and innumerable courses of study that explore the interstices between traditional disciplines (see Box 6-9 and Appendix D).

The administrations of many campuses have begun to respond vigorously with renewed energy and innovative organizational structures. After several decades of experimentation, interdisciplinary centers, institutes, programs, and other structural mechanisms have proliferated on and adjacent to university campuses; indeed, these research units often outnumber traditional departments (see Figure 1-1 and Box 1-1). Despite frequent tensions over budgets, space, and intellectual turf, many of these centers and institutes are vibrant research and training environments. They do not supersede the departments but complement them, often generating new kinds of excitement.

---

[6]Feist, G. J. and Gorman, M. E. 1998. The Psychology of Science: Review and Integration of a Nascent Discipline. *Review of General Psychology* 2, no. 1:3-47; Simonton, D. K. 2004. *Creativity in Science: Chance, Logic, Genius, and Zeitgeist.* New York: Cambridge University Press; Simonton, D. K. 2003. Scientific Creativity as Constrained Stochastic Behavior: The Integration of Product, Person, and Process Perspectives. *Psychological Bulletin* 129, no. 4:475-94.

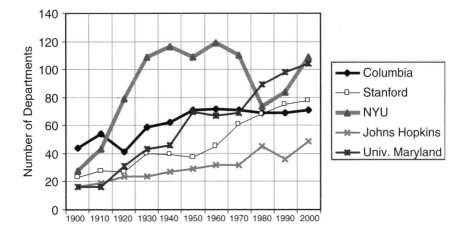

**FIGURE 1-1** Number of departments at selected universities, 1900-2000.
NOTES: The number of departments has increased steadily over the last century, from about 20 in 1900 to between 50 and 110 in 2000. National professional societies have also increased in number from 82 in 1900 to 367 in 1985 (see Figure 7-1). Although those changes may appear to indicate increasing specialization, the increases in new departments, such as biophysics and biochemistry, and societies, such as neuroscience and photonics, reflect a blending of previously distinct fields. SOURCE: The Committee was able to obtain department lists from small and large public and private institutions across the United States. NYU decreased after 1970 because the departments on their two campuses (University Heights and Washington Square) merged around that time.

## KEY CONDITIONS FOR INTERDISCIPLINARY WORK

During the preparation of this report, practitioners of IDR and other contributors described some of the key conditions for effective IDR. They include sustained and intense communication, talented leadership, appropriate reward and incentive mechanisms (including career and financial rewards), adequate time, seed funding for initial exploration, and willingness to support risky research (see Table 1-1).

## CONVERSATIONS, CONNECTIONS, COMBINATIONS

At the heart of interdisciplinarity is communication—the conversations, connections, and combinations that bring new insights to virtually every kind of scientist and engineer. Just as a biologist (Watson) and a

## STRUCTURE/POLICIES

### BOX 1-1 University Departments and Centers.
### Case Study: Columbia University

Columbia has been extremely supportive of interdisciplinary education and research, but it, like many other universities, has almost no publicly accessible records of the administrative structures used to facilitate such work.

Departments of instruction at Columbia are established by the trustees and written into the university statutes. Therefore, there are accurate records of their number. A list of current departments is published in the *Faculty Handbook*.[a] For historical information, prior handbooks are available in the archives.[b] Since 1950, department numbers have fluctuated (see Figure 1-1). Numbers alone, however, are not very enlightening. In each of the decades since 1950, some departments have been eliminated and replaced by others as the university shifted its academic priorities and some departments have been allowed to linger in the university statutes long after they cease to be functioning entities. A statutory count does not reveal how widely the university has dispersed its energies.

Unlike departments, centers, institutes, and other interdisciplinary units are not written into the university statutes. Institutes are supposed to require the approval of the university senate and the president. In contrast, centers and other interdisciplinary units can be created by the individual schools and in practice have often been established without even the approval of the dean. There is no central approval or recordkeeping. Lists of the interdisciplinary units were compiled for the university's last two accreditation reviews, in 1996 (105) and 2001 (241). In 2004, there were 277 such units. There are no counts for years prior to 1996.[c] Even more than academic departments, institutes and centers can vary substantially in size, resources, and contributions to the university. Some are bigger and intellectually more influential than some academic departments. Others are highly specialized and narrow. Some have existed for decades, others disappear after only a few years, and still others merge to create new units or emerge when one interdisciplinary unit is split. Some have retained their original purpose throughout their lifetimes; others have substantially shifted their academic focus. Aggregate numbers cannot reflect this diversity.

---

[a]The Columbia University Faculty Handbook is available on line at *http://www.columbia.edu/cu/vpaa/fhb/*.

[b]Columbiana Library Web Page *http://www.columbia.edu/cu/columbiana/collection.html*

[c]"We were not as systematic in our counting in 1995 as we were in later years and we, therefore, may have understated the number that actually existed in that year." Steven Rittenberg, Vice Provost, Columbia University, email communication, March 19, 2004.

**TABLE 1-1**   Key Conditions for Successful IDR at Academic Institutions Based on Committee Interviews with IDR Leaders and Scholars

| Aspect | Key Conditions |
|---|---|
| Initial Stages: Building Bridges | • Common problem(s) to solve<br>• Leadership<br>• Environment that encourages faculty/researcher collaboration<br>• Establishing a team philosophy<br>• Seed/glue money<br>• Seminars to foster bridges between students, postdoctoral scholars, and PIs at the same institution<br>• Workshops to foster bridges between investigators at different institutions<br>• Frequent meetings among team members<br>• Think of the end at the beginning |
| Supporting the Project | • Science and engineering PhDs trained in research administration<br>• Support project initiation and team building<br>• Seamless and flexible funding<br>• Willingness to take risks<br>• Recognize potential for high impact<br>• Involvement of funding organization |
| Facilities | • Physical co-location of researchers<br>• Shared instrumentation<br>• Enhance chance meetings between researchers, such as on-site cafeterias |
| Organization/ Administration | • Matrix organization<br>• Rewards for academic leaders who foster IDR<br>• Tenure/promotion policies for interdisciplinary work<br>• Utilize experts with breadth and IDR experience for assessment<br>• Professional recognition of successful practitioners of IDR |

physicist (Crick) a half-century ago enriched their insights with evidence from x-ray crystallography to imagine the structure of DNA, scientists in every research area are alert to flashes of understanding from other fields that may illuminate their own specialties. Without sustained and intense discussion of such possibilities and without special effort by researchers to learn the languages and cultures of participants in different traditions, the potential interdisciplinary research might not be realized and might have no lasting effect. Learning a new field is always hard work, and it must be catalyzed by both formal efforts, such as institutional policies that support new programs, and informal efforts, such as cafeterias, collaborative spaces, and common rooms that encourage mingling and conversation.[7]

The task of this report is to update and illuminate the intrinsic power of IDR and to build on models and recommendations that can identify and remove barriers to its full expression. A similar task has been assigned to research councils in Europe (see Boxes 1-2 and 1-3).

The purpose and current research agenda of IDR must be examined more closely than they have been by scholars. Should we be moving from a gradual trend toward interdisciplinarity to one that is even stronger? What is the proper response to the many knowledgeable observers who continue to advise "staying in one's long-cultivated disciplinary garden" as "the best way to produce the fruits of scientific discovery"?[8] In seeking the best ways to support research, policy makers must address difficult institutional, fiscal, and behavioral issues; they must also find better ways to assess the effectiveness of different research and teaching settings.

## A QUESTION OF URGENCY

Much depends on the nation's response to the challenges described in this report. Strengthening IDR is not merely a concept that is philosophically attractive or that serves the special interests of a few neglected fields. It has been vital since the creation of our great research universities—and critical during times of national emergency. It has led to major new industries and opened up the world to the creation of wealth, to international collaboration, and to enhanced technology and scientific exchange.

---

**Convocation Quote**

There is this long-standing call for this type of research. The question we have to ask ourselves is, what is the problem? Why isn't this proceeding at a more rapid rate?

Cliff Gabriel, deputy associate director,
White House Office of Science and Technology

---

[7]Participants in the committee's Convocation on Facilitating Interdisciplinary Research (see Appendix C) emphasized the importance of a conscious strategy to promote informal communication.

[8]Feller, I., Whither Interdisciplinarity (In an Era of Strategic Planning)? Presented at AAAS Annual Meeting, Seattle, WA, February 15, 2004.

STRUCTURE/POLICIES

## BOX 1-2 (1+1) > 2: Promoting Multidisciplinary Research[a] in the Netherlands

In 2002, the Dutch ministers of education, culture, and science (OC&W) and the minister of economic affairs (EZ) jointly asked the Dutch Advisory Council for Science and Technology Policy (AWT) for advice on how to foster multidisciplinary research.

The council's recommendations, published in 2003, are based on the central observation that multidisciplinary research is growing in importance. Scientific discoveries occur increasingly on the borders between disciplines. In addition, economic and social innovations call for input from a variety of disciplines.

In its recommendations, the council focuses on universities. It found that the obstacles to multidisciplinary research manifest themselves most emphatically in such institutions, which, paradoxically, are best positioned to gain from IDR. Universities play a key role in the knowledge infrastructure. After all, many of the students and research assistants trained at universities are the future "producers" and "consumers" of the results of research.

The recommendations are practical and grouped along three issues that, according to the AWT, are indispensable for the effective promotion of multidisciplinary research:

• Ensure that there are enough motivated researchers. Incentives are required to encourage scientists to engage in multidisciplinary research. In this connection, the council makes statements about a variety of subjects, including the desired broader definition of scientific quality, the broadening of university career policy, and the need to improve the image of multidisciplinary research.
• Promote interaction and meetings. Tangible measures are required to put this into practice. The council calls for the creation of more horizontal ties at universities and for the establishment of institutions to lead research in societal issues.
• Set challenging goals. Multidisciplinary research can be successful only if the goal, question, or ambition is attractive and shared. In this context, the council believes that it is essential to ensure that all the relevant disciplines are involved from the beginning. The council also presents concrete tools for achieving that.

In addition to the universities, the recommendations address the Ministries of OC&W and EZ, the Netherlands Organization for Scientific Research, and the Royal Netherlands Academy of Arts and Sciences.

---

[a]Report 54. (1+1) > 2. Promoting Multidisciplinary Research. September 2003. Advisory Committee for Science and Technology Policy (AWT). Available on the AWT home page *http://www.awt.nl/en/index.html*. Although the term multidisciplinary is used in the Netherlands, its definition fits the committee's definition of interdisciplinary (see Chapter 2).

STRUCTURE/POLICIES

## BOX 1-3   Interdisciplinary Research in Europe: The EURAB Report[a]

The European Union's research advisory board (EURAB)[b] released a report in April 2004 detailing the barriers to carrying out IDR in Europe and making recommendations as to how such barriers can be overcome.

EURAB found that barriers to IDR are highest where the traditional one-department, one-discipline structure of most universities is reflected in the structures of research funding bodies. Specific challenges include the difficulty of creating new interdisciplinary programs by using established one-discipline funding systems, the weakness of multidisciplinary career structures, the lack of established interdisciplinary scientific journals, and education systems that are not geared toward producing multidisciplinary graduates and postgraduates.

EURAB recommendations focused on a reassessment of disciplinary demarcations, a removal of structural and administrative barriers in and between institutions, and a rethinking of associated research training.

The report suggests that a reduction of the number of de facto definitions by which research funding is allocated would be helpful in creating greater opportunities for interdisciplinarity. EURAB cautioned against the unwitting creation of barriers to IDR when EU expert groups or advisory boards are being created.

With regard to the education and training of researchers, the report notes a need to provide bridges between disciplines at the undergraduate level and warns that overspecialization at the doctoral level creates barriers to industrial employment. EURAB recommended establishing a high-level EU interdisciplinary doctoral program and encouraged universities to provide opportunities for undergraduates to take credit modules outside their own specialties.

With regard to creating new IDR centers, EURAB recommended examining the advantages of virtual centers. When a new structure is proposed, the cost and benefits should be evaluated against the reform or extension of existing traditional disciplinary structures. EURAB recommended that any new center integrate teaching and research activities of traditional disciplinary departments.

Finally, with regard to research funding agencies, EURAB recommended transparent mechanisms to review interdisciplinary proposals, which may include flexible allocation to discipline-based review panels with cross-referencing and joint evaluation. In addition, EURAB requested a review of mechanisms that are used by EU and national funding agencies to design, evaluate, and manage IDR.

---

[a]Interdisciplinarity in Research, EURAB, April 2004. Available on line at *http://europa.eu. int/comm/research/eurab/pdf/eurab_04_009_interdisciplinarity_research_final.pdf.*

[b]European Research Advisory Board (EURAB) home page *http://europa.eu.int/comm/ research/eurab/index_en.html.*

To hinder this activity is to diminish our ability to address the great questions of science and to hesitate before the scientific and societal challenges of our time. If a disjunction exists between how science naturally moves and how various structures hold it back, the task is to mend it.

"I'M ON THE VERGE OF A MAJOR BREAKTHROUGH, BUT I'M ALSO AT THAT POINT WHERE CHEMISTRY LEAVES OFF AND PHYSICS BEGINS, SO I'LL HAVE TO DROP THE WHOLE THING."

The literature that this committee has reviewed suggests an evolution in modern research toward greater complexity. If that is valid, researchers need organizational and career structures that are suitably flexible and carefully designed to support the trend.

# 2

# The Drivers of
# Interdisciplinary Research

No one can predict the issues that science and society will consider most pressing in the decades to come. But if we look at some high-priority issues of today—such as world hunger, biomedical ethics, sustainable resources, homeland security, and child development and learning—and pressing research questions, such as the evolution of virulence in pathogens and the relationship between biodiversity and ecosystem functions, we can predict that those of the future will be so complex as to require insights from multiple disciplines. What research strategies are needed to address such a future? To what extent will interdisciplinary research (IDR) and interdisciplinary education be among the strategies? Just what is IDR?

## DEFINING INTERDISCIPLINARY RESEARCH

No single definition is likely to encompass the diverse range of activities that have been described under the heading of IDR. Reflecting the diversity of modes of interdisciplinary work, several organizational models have evolved (see Table 2-1). For the purpose of this report, the committee has developed the following description as a point of departure:

> Interdisciplinary research (IDR) is a mode of research by teams or individuals that integrates information, data, techniques, tools, perspectives, concepts, and/or theories from two or more disciplines or bodies of specialized knowledge to advance fundamental understanding or to solve problems whose solutions are beyond the scope of a single discipline or field of research practice.

Research is truly interdisciplinary when it is not just pasting two disciplines together to create one product but rather is an integration and synthesis of ideas and methods. An example is the current exploration of string theory by theoretical physicists and mathematicians, in which the questions posed have brought fundamental new insights both to mathematicians and to physicists.

---

**Convocation Quote**

Interdisciplinary research by definition requires the researchers to learn the other discipline. I like to stress vocabulary, but also methodology; I feel very strongly about it.

Ruzena Bajcsy, director of the Center for Information Technology Research in the Interest of Society, University of California, Berkeley

---

Other terms used include borrowing and multidisciplinary research.

- *Borrowing* describes the use of one discipline's methods, skills, or theories in a different discipline. A borrowed technique may be assimilated so completely that it is no longer considered foreign, and it may transform practice without being considered interdisciplinary.[1] An example of borrowing is the use of physical-science methods in biologic research, such as electron microscopy, x-ray crystallography, and spectroscopy. Such borrowing may be so extensive that the origin of the technique is obscured.[2]

- For purposes of this discussion, *multidisciplinary research* is taken to mean research that involves more than a single discipline in which each discipline makes a separate contribution. Investigators may share facilities and research approaches while working separately on distinct aspects of a problem.[3] For example, an archaeological program might require the participation of a geologist in a role that is primarily supportive. Multidisciplinary

---

[1]Klein, J. T. "A Conceptual Vocabulary of Interdisciplinary Science." *Practising Interdisciplinarity.* Eds. Weingart, P. and Stehr, N. University of Toronto Press, Toronto, 2000. pp. 3-24.

[2]See Holton, G., Chang, H., and Jurkowitz, E. "How a scientific discovery is made: A case history." *American Scientist*, Vol. 84, July-August 1996, pp. 364-75, for specific examples of borrowing.

[3]Friedman, R. S. and Friedman, R. C. "Organized Research Units of Academe Revisited." In *Managing High Technology: An Interdisciplinary Perspective.* Eds., Mar, B. W., Newell, W. T. and Saxberg, B. O. Amsterdam: North Holland-Elsevier, 1985. pp. 75-91.

STRUCTURE/POLICIES

**TABLE 2-1**   Interdisciplinary Research Structures

As a direct response to one component of its charge, "Identify and analyze current structural models of interdisciplinary research," the committee collected information on about 100 existing IDR groups and centers. The committee tested the categorization proposed by Epton et al.[a] and found that, although it is largely applicable, there are important additional IDR structural categories and characteristics, including national labs, space allocation, and fluidity of teams.

SMALL ACADEMIC (< 10 persons)
- Bottom-up initiation
- Research is primary; training is byproduct
- Loose management structure
- Many participants have disciplinary research commitments as well

LARGE ACADEMIC
- Bottom-up initiation, top-down incubation and management
- Research and training components
- Management by directors who report directly to vice president for research or equivalent
- Tend to be permanent features: new building, instrumentation
- Some centers "co-hire" faculty, but faculty are affiliated with departments
- Space allocation: mix of permanent and "hotel" facilities

INDUSTRY
- Top-down, product-driven research
- Focused on research, not training
- Structured management
- Discrete timelines and end points
- Fluid movement of researchers between teams

NATIONAL LABORATORIES
- Blend of top-down, mission-driven research and bottom-up initiation
- Research and training components
- Structured management
- Discrete timelines and end points
- Fluid movement of researchers between teams

INTERINDUSTRY, INTERUNIVERSITY, UNIVERSITY-INDUSTRY
- Top-down, societal needs-driven research (can be basic and applied)
- Research and training components
- Part-time directors with advisory boards
- Often initiated with large starting grants (such as National Science Foundation–funded Science and Technology Centers and Engineering Research Centers)
- Except for seed grants, faculty must provide own grant money
- Programs may offer an "immersion" IDR opportunity

[a]Epton, S. R., Payne, R. L., and Pearson, A. W. (1985) "Contextual Issues in Managing Cross-Disciplinary Research." In *Managing High Technology: An Interdisciplinary Perspective*. Eds. Mar, B. W., Newell, W. T. and Saxberg, B. O. New York: Elsevier. pp. 209-29.

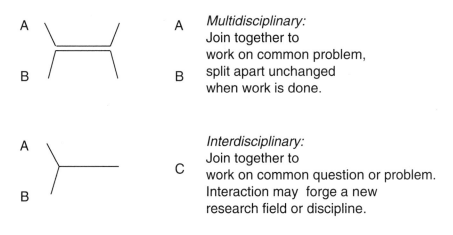

FIGURE 2-1   Difference between multi- and interdisciplinary.
SOURCE: Adapted from L. Tabak, Director, NINDS, NIH. Presentation at Convocation on Facilitating Interdisciplinary Research, Washington, D.C., January 29, 2004.

research often refers to efforts that are additive but not necessarily integrative (see Figure 2-1).[4,5]

IDR can also be described in terms of modes of participation. In one mode an individual investigator masters and integrates several fields. The investigator may conceive a new problem or method or may venture far enough from his or her original discipline to create a new field. For example, Albert Einstein ventured from his field of physics into Riemann geometry to describe his new General Theory of Relativity.

In a second mode, a group of investigators, each with mastery in one field, learn to communicate and collaborate on a single problem.[6] In some cases, such groups may be quite large, as in high-energy physics and genomics research.

---

[4]Porter, A. L. and Rossini, F. A. "Multiskill Research," *Knowledge: Creation, Diffusion, Utilization*, Vol. 7, No. 3, March 1986, p. 219.

[5]Klein, J. T. *Interdisciplinarity: History, Theory, and Practice*. Detroit: Wayne State University Press, 1990, p. 56.

[6]In one formulation, this mode is termed consilience: the "jumping together of knowledge" across disciplines "to create a common groundwork of explanation". Wilson, E. O. *Consilience: The Unity of Science*, New York: Alfred A. Knopf, 1998, p. 8.

**Convocation Quote**

If you think of disciplines as organs, true interdisciplinarity is something like blood. It flows. It is a liquid. It is not contained. There is no inside and outside.

Alice Gottlieb, professor of medicine and director, Clinical Research Center at the Robert Wood Johnson Medical School

The committee paid special attention to interdisciplinary *education*, viewing it as a central component of IDR. Students are prepared for the complexities of IDR when they are encouraged to understand and pursue multiple disciplines and to address complex problems from the perspective of multiple fields in their undergraduate and graduate studies. Specific suggestions for strengthening interdisciplinary education are presented in Chapters 4, 5, and 9.

## CHALLENGES DRIVING INTERDISCIPLINARY RESEARCH

To understand the natural world, scientists are drawn toward the unknown, especially toward the "grand challenges" of research. How did the universe originate? What physical processes control climate? What is the carrying capacity of the biosphere? Such challenges almost always invite journeys across disciplinary frontiers. A scientist may respond to many kinds of motivation, or "drivers," in undertaking interdisciplinary projects. We list four such drivers below, providing examples and exploring why the practice of modern science and engineering requires interdisciplinary work.

### The Inherent Complexity of Nature and Society

Human society in its natural setting contends with enormously complex systems that are influenced by myriad forces. It is not possible to study the earth's climate, for example, without considering the oceans, rivers, sea ice, atmospheric constituents, solar radiation, transport processes, land-use, land-cover, and other anthropogenic practices and the feedback mechanisms that link this "system of subsystems" across scales of space and time. A full predictive or even descriptive understanding requires the use of many disciplines (see Box 2-1).

Nature's complexity often leads to surprises that require much thought and experimentation to unravel. An example is the unexpected emergence of the Antarctic ozone hole in the austral springtime, a phenomenon found to be the consequence of complex chemical and dynamic pathways attributable to the use of chlorine- and bromine-bearing compounds in commercial

## EVOLUTION

### BOX 2-1 The International Geosphere-Biosphere Program (IGBP)

The real connections that link the geosphere and biosphere to each other are subtle, complex, and often synergistic; their study transcends the bounds of specialized, scientific disciplines and the scope of limited, national scientific endeavors. For these reasons progress in fundamental areas of ocean-atmosphere interactions, biogeochemical cycles, and solar-terrestrial relationships has come far more slowly than in specialized fields, in spite of the obvious practical importance of such studies. If, however, we could launch a cooperative interdisciplinary program in the earth sciences, on an international scale, we might hope to take a major step toward revealing the physical, chemical, and biological workings of the Sun-Earth system and the mysteries of the origins and survival of life in the biosphere. The concept of an International Geosphere-Biosphere Program (IGBP), as outlined in this report, calls for this sort of bold "holistic" venture in organized research—the study of whole systems of interdisciplinary science in an effort to understand global changes in the terrestrial environment and its living systems.[a]

So begins the preface to a 1983 workshop report that would help to launch the IGBP, which 20 years later is one of the largest interdisciplinary international research efforts ever undertaken. In its origins, the program reflected all the major drivers of IDR. It begins with the *complexity of nature*, the interactions between the land mass, the oceans of air and water, and the life forms of Earth. It finds that much of the most exciting science takes place on the boundaries of both systems and disciplines, such as the biogeochemical flows of the major life-support elements. Encouraging such explorations are powerful *societal needs* to understand how humankind is transforming the earth and the threats and opportunities that such transformation poses. Making possible such ambition are *generative technologies*, particularly computer simulation and modeling, remote sensing from space, and recovering the past from cores of ocean bottom, ice, lakes, and trees.

In scale, the program reflects both big science and local investigation. Some 10,000 scientists in 80 countries and more than 20 disciplines take part in IGBP scientific activities.[b] They include agricultural scientists, archaeologists, atmospheric chemists, and dynamicists, biologists, climatologists, ecologists, economists, environmental historians, geographers, geologists, hydrologists, mathematicians, meteorologists, plant physiologists, political scientists, physical and chemical oceanographers, remote sensing scientists, and sociologists.

The program has transformed the disciplines initially involved. Disciplines that were primarily focused on local and small scales, such as ecology, now address large-scale processes and conduct extensive experiments including in situ carbon enrichment and experimental deforestation. Disciplines that were primarily curiosity-driven such as the many paleosciences, have acquired important societal relevance. Natural and social sciences have come to need and value each other.

*continues*

[a]Friedman, H. Preface. *Toward an International Geosphere-Biosphere Program: A Study of Global Change*. Report of a National Research Council Workshop, Woods Hole, MA July 25-29, 1983. Washington: National Academy Press, p. vii.

[b]For these details and insights we are grateful to Will Steffen, Executive Director of the IGBP.

---

**BOX 2-1  Continued**

Disciplines have discovered common interests, such as how to relate wholes to parts, macro processes to micro behavior, and global to local. Indeed, global change science now exhibits many interdisciplinary aspects, with a second generation of scientists transcending their disciplines and schooled in problem-driven common knowledge.

But most important are the major scientific findings. The program has transformed our understanding of both nature and humankind. A recent summary volume[c] finds that:

- The earth is a system that life itself helps to modulate. Biological processes interact with chemical and physical processes to create the planetary environment.
- Human activities are influencing the functioning of the earth system in many important ways.
- The earth is operating in a no-analogue state. The magnitudes and rates of changes occurring simultaneously in the earth system are unprecedented.
- The earth's dynamics are characterized by critical thresholds and abrupt changes.

---

[c]Steffen, W., Sanderson, A., Tyson, P., Jäger, J., Matson, P., Moore III, B., Oldfield, F., Richardson, K., Schellnhuber, H-J., Turner II, B. L., Wasson, R. *Global Change and the Earth System: A Planet Under Pressure*. IGBP Global Change Series.New York: Springer-Verlag, Berlin Heidelburg, 2004, 336 pp.

---

products. Pinpointing that cause required the combined efforts of many scientific and technical disciplines; solving the problem itself required the collaboration of physical scientists, engineers, economists, and social scientists.

Similarly, the human-genome mapping project was a complex undertaking that depended on extensive collaboration across many fields, including the biological and computational sciences. Basic questions of life—how living beings grow, how the brain functions, why many animals need to sleep, how retroviruses function—share the characteristic of complexity, and understanding them, even in part, depends on multiple disciplines. Gaining such understanding will almost certainly require deep expertise both at the subsystem level and at the interdisciplinary level—and the integration of these two levels. It is important to note that depth in research is not confined to single-discipline investigations. Statistical mechanics, for example, unites physicists and mathematicians in studies of substantial depth.[7]

If science and engineering deal with extremely complex systems, the same is true for studies of human society. How human societies evolve,

---

[7]Kafatos and Eisner, ibid. p. 1257.

make decisions, interact, and solve problems are all matters that call for diverse insights. Very fundamental questions are inherently complex. For example, why do humans kill each other? Why does hunger persist in a world of plenty? Answering such questions successfully requires collaboration across the natural sciences, social sciences, and humanities.

### The Drive to Explore Basic Research Problems at the Interfaces of Disciplines

Some of the most interesting scientific questions are found at the interfaces between disciplines and in the white spaces on organizational charts. Exploring such interfaces and interstices leads investigators beyond their own disciplines to invite the participation of researchers in adjacent or complementary fields and even to stimulate the development of a new interdisciplinary field. Examples include the following:

•  Biochemistry was long ago considered an interdisciplinary activity; today it has departmental, program, or similar structural status in most major universities.
•  The field of cognitive science has evolved in response to questions that could not be answered by single disciplines. Today the Cognitive Science Society embraces anthropology, artificial intelligence, neuroscience, education, linguistics, psychology, and philosophy.[8]
•  As biology has become more quantitative, its points of overlap with the mathematical sciences and the physical sciences have become more numerous and important. Today, the computational and statistical power of mathematics and the research facilities of the physical sciences are required for making sense of, for example, genomics, proteomics, epidemiology, structural biology, and ecology.
•  Ecology and economics (and other social sciences) have a common origin, at least in name, and, increasingly, a common field—ecologic economics—that aspires to facilitate "understanding between economists and ecologists and the integration of their thinking" with the goal of developing a sustainable world.[9]

That many of the most interesting scientific questions are lodged in the interstices between disciplines can also be seen in various activities that honor outstanding creativity. For example, although the MacArthur Foundation fellow awards are not given on the basis of interdisciplinarity, to

---

[8]See Appendix D on the development of disciplinary societies.
[9]The Web site of the International Society of Ecological Economics is *http://www.ecologicaleconomics.org*.

judge from brief biographies, two-thirds to three-fourths of MacArthur fellows in science appear to work in interdisciplinary fields.

### The Need to Solve Societal Problems

Human society depends more than ever on sound science for sound decision making. The fabric of modern life—its food, water, security, jobs, energy, and transportation—is held together largely by techniques and tools of science and technology. But the application of technologies to enhance the quality of life can itself create problems that require technological solutions. Examples include the buildup of greenhouse gases (hence global warming), the use of artificial fertilizers (water pollution and eutrophication), nuclear-power generation (radioactive waste), and automotive transportation (highway deaths, urban sprawl, and air pollution).

---

### EVOLUTION

### BOX 2-2   The Development of Microwave Radar at MIT's Radiation Laboratory[a]

The development of radar (radio detection and ranging) during the 1940s was largely accelerated by military needs in World War II. Members of the scientific community recognized the value of radar to the war effort. In the United States, the effort to expand microwave radar capabilities was concentrated at MIT's Radiation Laboratory, which was staffed by civilian and academic scientists in many disciplines. Projects included physical electronics, microwave physics, electromagnetic properties of matter, and microwave communication principles.

The "Rad Lab" was responsible for almost half the radar deployed in World War II and at one point employed almost 4,000 people working on several continents in government, industrial, and university laboratories. What began as a British-American effort to make microwave radar work evolved into a centralized laboratory committed to understanding the theories behind experimental radar while solving its engineering problems.

The Rad Lab was formally shut down after the end of World War II in 1945, but in 1946 the Basic Research Division was incorporated into the new Research Laboratory of Electronics at MIT. Research continued on problems in physical electronics and microwave physics. Modern techniques were applied to physics and engineering research, and engineering applications were emphasized in microwave communication.

---

[a]MIT Radiation Laboratory series Volume 28. Ed. Henney, K. Available at *http://www. brewbooks.com/ref/rl/ref_radlab_v28.html*; G.Goebel, Microwave Radar & The MIT Rad Lab. Available at *http://www.vectorsite.net/ttwiz3.html*; Lab's Microwave Traditions at RLE. RLE currents, Vol. 4, No. 2—Spring 1991. Available at *http://rleweb.mit.edu/radlab/radlab.HTM*.

An indication of interdisciplinarity in response to societal needs is the success of large, sustained endeavors, many of which continue to this day. During World War II, for example, science and engineering demonstrated the ability to strengthen military power rapidly (see Box 2-2). The 3-year Manhattan Project (1942-1945) to develop an atomic bomb was an interdisciplinary effort requiring researchers from many fields and subfields of science and engineering, from the wide sweep of chemistry and physics to the specific skills of uranium refinement, isotope separation, plutonium purification, nuclear decay measurement, nuclear-waste disposal, and radiation biology.

Another example is the National Cancer Act, signed by President Nixon in 1971. The act authorized an interdisciplinary research effort involving a vast sweep of biomedical disciplines, from genetics and cell biology through clinical care, bioethics, and biostatistics. Cancer research has always been among the most interdisciplinary of fields, mirroring the complexity of the many diseases it addresses.

Researchers continue to apply the 20th century's revolutionary genetic insights to unravel the structures and functions of proteins (see Box 2-3). This investigation influences every aspect of the life sciences, at every level, from molecular arrangements to clinical, population, and ecologic studies.[10]

## The Stimulus of Generative Technologies

Generative technologies are those whose novelty and power not only find applications of great value but also have the capacity to transform existing disciplines and generate new ones. An early momentous example was the use of microscopes by Hooke and van Leeuwenhoek to view "cubicles," or cells, in animal and plant bodies and to make it possible to see living "animalcules" (bacteria) with their own eyes—both critical steps along the path to modern molecular biology.

A recent example of a generative technology has been the development of the Internet, whose popular form is only about 10 years old. The Internet

---

[10]Yet another example can be found in Branscomb, L., Holton, G., and Sonnert, G. *Cutting-edge Basic Research in the Service of Public Objectives: A Blueprint for an Intellectually Bold and Socially Beneficial Science Policy.* Consortium for Science Policy Outcomes, Arizona State University, May 2001. Available on-line at *http://www.cspo.org/products/reports/scienceforsociety.pdf* (Based on a workshop sponsored by the David and Lucile Packard Foundation and the Alfred P. Sloan Foundation.) The report makes the case for use-inspired or "Jeffersonian" basic research and includes a master list of questions in science and technology, most of which require interdisciplinary approaches. Holton, G., "What Kinds of Science are Worth Supporting?" The Great Ideas Today, Encyclopedia Britannica, Chicago, 1998.

---

EVOLUTION

**BOX 2-3   Protein Structure Determination Using X-Ray Crystallography[a,b]**

The knowledge of protein structures is critical to fighting disease with drugs. In recent years, the development of new techniques to determine protein structure, combined with rapid improvement in computer technology, has allowed protein-structure determination to proceed at a rate that is keeping pace with advances in biomedical science. In the case of x-ray crystallography, its development and wide use in protein-structure determination–which spanned a century–began with no knowledge of its value for biomedicine.

X rays were first discovered in 1895, and the diffraction of x rays by electrons in crystals was first demonstrated in 1912. In the 1930s, x rays were aimed at crystals of biological molecules, but it was not until Perutz and Kendrew determined the molecular structure of hemoglobin and myoglobin in 1960 that the value of x-ray crystallography in protein science was realized. In the 1970s, synchrotron radiation (see Box 2-5) was harnessed as a source of x rays for protein crystallography, and the 1990s saw a great increase in the number of protein structures determined with this technique. Research to develop the technology was an interdisciplinary endeavor. Its long-term nature should remind those who facilitate IDR that support of basic research can often have payoffs that are not immediately visible and are often outside the field in which they were initially envisioned.

---

[a]Dill, K. Strengthening Biomedicine's Roots. *Nature* 22 400:309-310. July 1999.
[b]History of X-ray Crystallography and Associated Topics. Available at *http://www.dl.ac.uk/SRS/PX/history/history.html.*

---

has both enhanced connectivity between people and revolutionized access to information, transforming the ability to interact and collaborate across space and time. It has special relevance to the world of research, for which it offers ways to work in large, distributed teams, enlarge the educational enterprise, provide access to data on time and spatial scales never possible before, and design powerful new tools to transform the processes of discovery, learning, and communication (see Box 2-4).

Dramatic declines in the cost of processing, storing, and transmitting information are transforming science and engineering disciplines. Some experts have called on the National Science Foundation and other science agencies to launch a bold new initiative in cyberinfrastructure, which would play the same role in supporting the knowledge economy that roads, power grids, and rail lines have played in supporting the industrial economy.[11]

---

[11]Revolutionizing Science and Engineering through Cyberinfrastructure. Report of the National Science Foundation Advisory Panel on Cyberinfrastructure. February 3, 2003. Available at *http://www.cise.nsf.gov/sci/reports/toc.cfm.*

---

INNOVATIVE PRACTICE

**BOX 2-4 The Knowledge and Distributed Intelligence (KDI) Funding Initiative**

The rise in computer power and connectivity is reshaping relationships among people and organizations and transforming the processes of discovery, learning, and communication. The knowledge and distributed intelligence (KDI) funding initiative at the National Science Foundation (NSF) was created in 1998 to find ways to model and make use of complex and cross-disciplinary scientific data.[a] KDI supported interdisciplinary projects of individuals or groups that took advantage of changes in how research was being done, such as increases in computing power and connectivity among researchers. The initial solicitation had three foci of research: knowledge networking, learning and intelligent systems, and new computational challenges. The KDI initiative has sponsored research that analyzes living and engineered systems in new ways, and it encourages investigators to explore the cognitive, ethical, educational, legal, and social implications of new types of learning, knowledge, and interactivity.

A program assessment was carried out in 2002. NSF recognized that metrics have to be developed that match the goals of the research program. To that end, KDI grantees were invited to a workshop to determine how projects were organized and managed, to identify the projects, outcomes, and to catalog suggestions that might help future grantees in their execution of KDI-sponsored projects.

The evaluation[b,c] provides interesting information about tools, research directions, outreach, and student training. Management of collaborative and multidisciplinary research projects was a substantive issue. Project success depended largely on coordinating interactions among researchers. Dispersion of participants, rather than interdisciplinarity, was the most problematic aspect of KDI projects. Projects with principal investigators in multiple universities were substantially less well coordinated and reported fewer favorable outcomes. Project-related conferences, workshops, and other regular meetings appeared to reduce the adverse effects of dispersion. The assessment identified a number of needs for further support, including management tools that would increase the ease with which project participants interact over the lifetime of the project.

---

[a]The original KDI solicitation is available at: *http://www.nsf.gov/pubs/1998/nsf9855/nsf9855.pdf.*

[b]Cummings, J. and Kiesler, S. (2004) KDI Initiative: Multidisciplinary Scientific Collaborations. NSF Report. Available on the NSF KDI Home Page: *http://www.cise.nsf.gov/kdi/about.html.*

[c]Taking stock of the KDI: Science of Evaluation. *http://www.cise.nsf.gov/kdi/eval.html.*

---

This cyberinfrastructure might be composed of distributed, high-performance computers, online scientific instruments and sensor arrays, multidisciplinary collections of scientific data, software toolkits for modeling and interactive visualization, and tools that enable close collaboration by physically distributed teams of researchers (see Box 2-5 and Box 9-7).

EVOLUTION

**BOX 2-5  Tool-Driven Interdisciplinary Research:
The Advanced Photon Source (APS) at
Argonne National Laboratory**

The Advanced Photon Source (APS) at Argonne National Laboratory is a national synchrotron-radiation light-source research facility. Commissioned in 1995, the APS is funded by the US Department of Energy, Office of Science, Office of Basic Energy Sciences.[a] Members of the international research community use high-brilliance x-ray beams from the APS to carry out basic and applied research in materials science; biology; physics; chemistry; environmental, geophysical, and planetary science; and innovative x-ray instrumentation.

Researchers come to the APS as members of collaborative access teams (CATs) or as independent investigators. CATs comprise large numbers of investigators with common research objectives and are responsible for design, construction, funding, and operation of beamlines at the facility. CATs must allocate 25 percent of their x-ray beam time to independent investigators or groups not affiliated with CATs.

The APS was designed to accommodate up to 32 CATs, of which over 20 are in operation. One of the interdisciplinary industry-university collaborations established to take advantage of APS resources is the University of Michigan-Howard University-AT&T Bell Laboratories (MHATT) CAT, formed in 1989. MHATT-CAT studies range from basic protein dynamics to the behavior of solid-state lasers. According to one of the directors of the MHATT-CAT, University of Michigan Physics Professor Roy Clarke, "a very important part of the project is to establish high-speed communications that link participating institutions and the facility at the APS, so that our students, particularly our undergraduate researchers, can participate actively in the research while attending classes on their respective campuses."[b]

Others CATs are run by university or industry teams. To enhance communication among and between teams, the APS Web site provides a linked list of CATs and offers a listserver for inter-CAT communication. The APS Web site also lists meetings of interest to facility users and highlights recent research by posting abstracts and figures on its home page.

---

[a]APS: Advanced Photon Source at ANL. Home page: *http://epics.aps.anl.gov/aps.php.*
[b]Elgass, J .R. (1994) Clarke co-directs project at Argonne photon facility. The University Record. University of Michigan. March 28, 1994. Accessed March 29, 2004 at *http://www. umich.edu/~urecord/9394/Mar28_94/2.htm.*

---

Advocates of cyberinfrastructure believe that it will allow a growing number of researchers to collect, process, analyze, and make available volumes of information that trigger shifts in the kinds of scientific questions that can be pursued; simulate systems of greater complexity and importance; and more easily work across scientific disciplines. For example, the National Science Foundation has funded a "National Virtual Observatory"[12] that is likely to transform astronomy. Within a few years, comprehensive sky surveys will be generating petabytes (quadrillions of bytes) of data every year. The long-term goal is to make this data available to every researcher, along with the databases, data mining algorithms, and visualization tools needed to make sense of it. Researchers believe that this information abundance will lead to qualitatively new science, such as statistical astronomy that analyzes the large-scale structure of the universe, and automated searches for exotic or previously unknown types of celestial objects.

Magnetic resonance imaging (MRI) is another example of a generative technology. The Nobel Prize in medicine and physiology for 2003 was awarded to chemists Paul Lauterbur and Peter Mansfield to honor their work that led to MRI. Their research grew out of a fundamental interest in using the magnetic resonance effect to produce images in proton-containing matter. MRI and positron-emission tomography (PET), with ancillary mathematical advances in tomographic analysis, have revolutionized many aspects of medical diagnosis and opened opportunities for safe experimentation with human subjects in the cognitive sciences.[13]

## CONCLUSIONS

The potential power of IDR to produce novel and even revolutionary insights is generally accepted. Ultimately, however, the value of IDR to the scientific enterprise depends on the extent to which individual researchers are free to engage in it. IDR must be not only possible but also attractive for students, postdoctoral fellows, and faculty members.

## FINDINGS

Interdisciplinary research (IDR) is a mode of research by teams or individuals that integrates information, data, techniques, tools, perspectives, concepts, and/or theories from two or more disciplines or bodies of specialized knowledge to advance fundamental understanding or to solve problems whose solutions are beyond the scope of a single discipline or area of research practice.

---

[12]US National Virtual Observatory. *http://www.us-vo.org/*.

[13]In one view, "new technologies are now driving scientific advances as much as the other way around. These technologies are enabling novel approaches to old questions and are posing brand-new ones." Leshner, A. I. "Science at the leading edge," *Science* Vol. 303:729. Feb. 6, 2004.

"I CAN REMEMBER WHEN ALL WE NEEDED WAS SOMEONE WHO COULD CARVE AND SOMEONE WHO COULD SEW."

IDR is pluralistic in method and focus. It may be conducted by individuals or groups and may be driven by scientific curiosity or practical needs.

Interdisciplinary thinking is rapidly becoming an integral feature of research as a result of four powerful "drivers": the inherent complexity of nature and society, the desire to explore problems and questions that are not confined to a single discipline, the need to solve societal problems, and the power of new technologies.

Social-science research has not yet fully elucidated the complex social and intellectual processes that make for successful IDR. A deeper understanding of these processes will further enhance the prospects for creation and management of successful IDR programs.

# 3

# Interdisciplinarity in Industrial and National Laboratories

A lthough the major emphasis in this study is on the state of IDR in academic institutions, academic institutions make up only one part of a pluralistic research enterprise. Some large industrial and national laboratories, which constitute other elements of the enterprise, have deep traditions of interdisciplinary research (IDR), partly because their R&D strategies must be able to respond to complex problems or challenges that require expertise in multiple fields and technologies. For example, when most experiments or systems are being developed or constructed, there is no choice but to be interdisciplinary. Experimental work in a genetics laboratory is likely to involve biology, organic and inorganic chemistry, flow physics structures to hold pieces together, electric circuits and electrochemistry computation, etc. Top-down management structures allow for easy horizontal movement of researchers in response to skill needs. The challenge is the degree of professionalism and collaboration to be brought to a project that involves many disciplines, skills, professionals, students, and technicians that form the cooperating team for some or all the projects' life span.

Such nonacademic laboratories are essential to the national R&D enterprise for both their research and training functions in science and engineering. This chapter discusses a sampling of nonacademic practices that have assumed growing relevance as more research universities have developed ties with industrial and federal agencies.[1] Most of the few studies of nonacademic IDR were published several decades ago, before the recent and substantial changes in many practices, such as the down-sizing of industrial laboratories. This discussion is by necessity largely restricted to anecdotal information and examples that are intended to span a representative array of practices and settings.

Faculty members in many universities are increasingly involved in outside consulting, research partnerships, or entrepreneurial efforts of their own, and thorough knowledge of nonacademic practices can add value to their own careers.[2] In addition, most graduate students who acquire PhDs in science and engineering will find career opportunities in nonacademic research settings, where most of the new research positions are likely to be created over the next few decades.[3] For today's students—who may eventually work not only with researchers in different science and engineering fields but also in development, marketing, law, economics, ethics, or other non-research activities—it is doubly important to hone their skills in communicating with people in other fields and to gain exposure to IDR in nonacademic settings through cooperative programs, summer jobs, and other opportunities.

## RESEARCH STRATEGIES AT INDUSTRIAL LABORATORIES

The first formal industrial R&D programs in the United States were organized just over a century ago. In 1900, for example, General Electric began funding the General Electric Research Laboratory in Schenectady, New York, to generate and use scientific knowledge. The nation's adoption of industrial R&D was prompted partly by Americans' exposure to industrial practices in Germany (the GE laboratory was directed by the German emigré Charles Steinmetz) and elsewhere in Europe (see Box 3-1), which emphasized the value of industrial research and industrial support for university research and graduate training.

The greatest expansion of industrial research came during the years after World War II, when the largest industrial laboratories—notably DuPont's Experimental Station in Wilmington, Delaware; IBM's Watson

---

[1]For a discussion of the effects of recent changes on the "research-university complex," see Conn, R. "The Research University Complex in a New Era: An Inquiry and Implications for Its Relationship with Industry," Washington, D.C.: Government-University-Industry Research Roundtable, 1999.

[2]Frosch, R. "Research and development," *Encyclopedia of Applied Physics*, Vol 16, Hoboken, N.J.: VCH Publishers, Inc., 1996, p. 419.

[3]COSEPUP (Committee on Science, Engineering, and Public Policy), *Reshaping the Graduate Education of Scientists and Engineers*, Washington, D.C.: National Academy Press, 1995.

## INNOVATIVE PRACTICE
### BOX 3-1   Philips Physics Research Laboratory

An early example of industrial IDR was the Philips Physics Research Laboratory in the Netherlands, which adopted explicit interdisciplinary policies as long ago as the 1930s. R&D activities began with the lightbulb and expanded steadily toward new challenges as the possibility of products appeared: radio-receiver bulbs, then radios themselves, telephony systems, sound equipment, and a long history of S&T-driven electronic products, each of which required many skills to develop.

Gilles Holst, founder and first director of Philips, refused to divide his laboratories by discipline, arguing that a team working on, for example, magnetic ferrites should have not only physicists but also chemists, crystallographers, and electrical engineers. He also created a development process that involved back-and-forth communication between the central laboratory and small R&D operations in each of the factories. He promoted a laboratory culture in which both academic excellence and industrial excellence were stimulated, and corporate leadership acknowledged the industrial laboratories as indispensable in product diversification and new business activities.

Among Holst's laboratory-management principles were the following:

- Hire young, intelligent researchers who have some experience in scientific research.
- Do not overemphasize the specific details of the research they have done, but consider their overall abilities.
- Give researchers freedom and accept their individual peculiarities.
- Let them publish and participate in international scientific activities.
- Avoid over-stringent organization; allow authority to arise naturally out of competence.
- Organize the laboratory not according to different disciplines but by interdisciplinary teams.
- Allow freedom in the choice of research subjects, maintaining an awareness of company needs.
- In individual research projects, do not interfere in the details, and assign no budgets.
- Reassign skilled senior researchers from the laboratory to applied R&D.
- Let the choice of research projects be determined by the state of the art in scientific knowledge.

The success of Philips's approach can be seen both in specific outputs, such as the invention of the compact disk and its successor, the DVD, and in its continuing global competitiveness. It is one of the few consumer electronics companies that supports large and multidisciplinary R&D operations.

Research Center in Yorktown Heights, New York; AT&T's Bell Laboratories at Murray Hill, New Jersey, and Xerox's Palo Alto Research Center (PARC) in California—set global standards of excellence in problem-driven interdisciplinary research and development. By the end of the 20th century, industry was providing just over half the funding for the nation's R&D activities and the federal government just over 40 percent. Of the total R&D spending, just over one-fourth went to research and the rest to development—proportions that have been typical since World War II.[4]

Most centralized research laboratories experienced downsizing that began in the 1980s and a shift in emphasis from research toward development. Even so, industrial R&D has retained its interdisciplinary character and its inherent flexibility. Reasons for this according to experts cited and interviewed for this chapter include the hierarchical structure of industrial research; the more focused, less open-ended nature of its goals (for example, to produce a more effective vaccine or electronic display); and the lack of the kind of tenure system common in academe.

> Our work in Pfizer in discovering and developing new medicines is critically dependent on integrating advances in many other fields from physics, chemistry, materials sciences, and engineering to computer modeling and information technology. By sharing ideas from these fields, our scientists are able to create a critical intellectual mass that increases the creativity, the capacity, and the speed of innovation at Pfizer and other companies like us.
>
> William C. Steere, Jr., chairman of the board and chief executive officer, Pfizer, Inc.
>
> In Council on Competitiveness, Going Global: The New Shape of American Innovation, 1998, p. 6.

### Some Models and Lessons from Industry

Virtually all industrial laboratories incorporate multiple disciplines of science and engineering, but an even greater degree of interdisciplinarity may occur during times of particular challenge. The examples below show how interdisciplinarity has been extended beyond the laboratory to reach throughout the corporate setting and even into customer relationships in

---

[4]Hounshell, D. A. "The Evolution of Industrial Research in the United States" in *Engines of Innovation: U.S. Industrial Research at the End of an Era*, Boston, MA: Harvard Business School Press, 1996, pp. 13-15.

addressing the demands of global competition, shorter product cycles, and quickly shifting customer needs.

### The Joint Programs of IBM

IBM, like other research-based corporations, has always emphasized IDR (Box 3-2). During the 1980s, however, when its profitable lines of hardware in computing and telecommunications evolved into unprofitable commodities, the firm learned how quickly the value of a research portfolio can decline. IBM kept its emphasis on IDR, but added a mechanism to more quickly communicate vital market facts to its researchers. The company developed a series of programs in advanced technology and early product development that were jointly planned, staffed, and funded by the research division and the appropriate product divisions and laboratories. Thus, both research and development activities benefited from the input of those who manufacture and market the outputs of research. That approach was extended to projects jointly developed by researchers and customers in recognition that the customer knows best what is most useful. The relationship between research and manufacturing has deepened with the creation of a manufacturing-research group within the research division, a move credited with saving hundreds of millions of dollars a year.[5]

### The Reinvention of Xerox

The history of the Xerox Corporation has been described in numerous accounts, including John Dessauer's *My Years at Xerox: The Billions Nobody Wanted* (1971), which described the development of the xerographic technology that revolutionized office copying. Smith and Alexander's *Fumbling the Future* (1988) recounts Xerox PARC's invention of the paradigm that led to personal computing, client-server architecture, graphical user interfaces, local area networks, laser printing, bit maps, and other advances but brought Xerox almost no economic benefit. Indeed, the business decline of Xerox in the middle 1980s is a vivid example of how brilliant research may fail to support a corporation when results are not translated into product development, marketing, and sales.

In the late 1980s and early 1990s, corporate management recognized the lack of clarity about research's role and its integration into the total

---

[5]Armstrong, J. "Reinventing research at IBM," in *Engines of Innovation: U.S. Industrial Research at the End of an Era.* Eds. Rosenbloom, R. S. and Spencer, W. J. Boston, MA: Harvard Business School Press, 1996, pp. 151-4.

## BOX 3-2   The Role of IDR at IBM[1]

IDR has been an integral part of IBM for 24 years and has allowed us to differentiate IBM from its competitors. One reason IBM has been able to sustain its basic-research program and remains the only large industrial basic laboratory today is its commitment to interdisciplinary teams.

Technology does not move along a linear path, and we need to have an interdisciplinary team already in place when problems come up. For example, when we had early evidence that bipolar transistors would soon reach the end of their ability to scale, it took scientists and engineers from many disciplines to spot the trend and find an answer. The answer was complementary metal oxide semiconductor (CMOS) technology. Companies that do not see the importance of IDR may not survive when times are challenging or when it is time to fundamentally change the direction of a company.

For IDR to be successful, a company must:

• Have an executive management team that believes in IDR and makes it a fundamental part of the culture. At IBM a physical sciences "coffee" has been held for 50 years to encourage talk across disciplinary boundaries.

• Form teams that include diverse skill sets. No research program has failed because it was an IDR program. Failures occur because there is an insufficient mass of the skills needed for an activity, such as having only one electrical engineer on a team when six were needed.

• Maintain an inventory of the diverse skills in the company. IBM's skills inventory has allowed appropriate interdisciplinary teams to be assembled quickly when needed for an urgent new project. Over time this "skill-finder" function has been automated.

Some of the lessons drawn from IBM's experiences may hold relevance for academe:

• Stimulate more interaction across disciplinary lines. At IBM more "points" are given in the personnel review process to people who interact and communicate across disciplines.

• Provide an incentive and reward system that encourages joint authorship of papers with those in other departments.

• Fund mini-sabbaticals in which a faculty member joins another department for a half-year every 3.5 years to understand the culture and challenges of other departments and disciplines.

---

[1]From comments for the committee by Bernard S. Meyerson, IBM fellow, vice president, and chief technologist, IBM Systems and Technology Group.

business. Xerox's business was successfully reorganized around a single focus (the "document company"). One lesson from this history may be that interdisciplinary *research* alone is not always sufficient in an industrial setting. R&D activities must be integrated with the surrounding business, including manufacturing and marketing, if research results are to contribute to profitability.[6]

## Colocation at Intel[7]

Intel, which chose not to create a corporate research unit, instead immerses researchers in the environment of the production line in its own version of interdisciplinary practice. The company's strategy was to recruit talented PhDs and spread them throughout the organization. The production line then became a seamless extension of the research laboratory; this allowed researchers to see perturbations, introduce bypasses, add steps, and explore variations in existing technologies with great efficiency. The company tries not to change production processes dramatically, but when a promising direction appears, it can set up a separate organization to explore it.

The principle underlying the strategy is that of "minimum information," set out by Intel cofounder Robert Noyce, guessing the answer to a problem and developing it as far as possible in a heuristic way. If that does not solve the problem, one starts over and learns enough to try something else. Clues are gathered from manufacturing engineers and others along the production line and from university collaborators with appropriate research expertise. In addition, the company maintains a small IDR group charged with staying abreast of broad developments in the semiconductor industry.

## The "Skunkworks" Model

To counteract ingrained and nonproductive organizational patterns, the concept of the "skunkworks" was developed, first at Lockheed Martin, to give creative freedom to a small, hand-picked team that is geographically removed from the main physical plant. A skunkwork is a small, loosely structured corporate research and development unit or subsidiary formed

---

[6]Myers, M. "Research and change management in Xerox," in *Engines of Innovation*. Eds. Rosenbloom, R. S. and Spencer, W. J. Boston, MA: Harvard Business School Press, 1996, pp. 133-49.

[7]Moore, G. "Some Personal Perspective on Research in the Semiconductor Industry," in *Engines of Innovation*. Eds. Rosenbloom, R. S. and Spencer, W. J. Boston, MA: Harvard Business School Press, 1996, pp. 165-74.

to foster innovation. The objective of the skunkworks may be sharply defined in terms of goal and timing. Notable skunkworks successes have included the U-2 and WR-71 Blackbird high-altitude spy planes, IBM's first personal computer, and Steve Jobs's breakthrough Macintosh computer at Apple. In one account of a successful skunkworks program, management researchers reported delivery of multiple related projects in a coordinated sequence that minimized the material and person-year costs, met new-product time-to-market deadlines by constructing production facilities in record time, met or exceeded company industrial standards, and created and documented new procedures for future projects.[8]

The concept of removing a small group with special autonomy has been criticized for lowering morale among those who are left behind and perceived to be "less than special."[9] But such resentment is less likely to form when a learning history of the project is carefully documented and provides for the transfer of new system tools to the main research facility.[10] An apparent lesson is that the skunkworks IDR model needs to be carefully adapted to each new setting.[11]

## A New Degree of Interdisciplinarity?

Industry is expanding the character of IDR to address problems of global scale. Recently, a large, high-profile consortium was announced at Stanford University that not only is interdisciplinary but combines influential sponsors in widely different sectors of business: ExxonMobil, General Electric, Schlumberger, and Toyota. The 10-year, $225 million Global Climate and Energy Project (GCEP) will bring together leading scientists from universities, research institutions, and private industry to collaborate on fundamental precommercial research. The strategy is to intensify research on hydrogen and renewable energy, $CO_2$ capture and storage, combustion science, and other promising technologies with the objective of developing

---

[8]Bommer, M., DeLaPorte, R., and Higgins, J. "Skunkworks approach to project management," *Journal of Management in Engineering*, Vol. 18, No. 1, January 2002, pp. 21-28.

[9]Schrage, M. "What's that bad odor at innovation skunkworks?," *Fortune*, Vol. 140, Issue 12, December 20, 1999, p. 338. Schrage writes, "This kind of 'innovation apartheid' may occasionally give birth to great new ideas, but it almost always breeds even greater resentment. Smart, capable people hate being marginalized."

[10]Bommer et al., p. 28.

[11]For example in a variant of the skunkworks model, a company seeds a small group that forms a startup company to work on a problem of interest to the parent company; if successful, the small company is then bought by the parent company.

energy systems that have low greenhouse emissions and can be used on a global scale.

---

### Training PhDs for Interdisciplinarity

The training of new PhDs is too narrow, too campus-centered, and too long. . . . In my view, radical change is not required to improve the overall effectiveness of PhD-level training. Training by apprenticeship under the direction of an expert really does work: It provides both new research and training simultaneously. . . . We should explicitly encourage PhD students to spend time in 'user environments' outside the university as part of their apprenticeship—perhaps internships analogous to the co-op programs often used by undergraduate and master's degree students. The ultimate aim of these internships should be to provide technical work experience that is as unlike academic experience as possible. So, for some careers, internships in manufacturing are preferable to internships in a corporate research lab.

Industry can play a valuable role in planning for these internships. The willingness of firms to take on graduate students will depend on factors that vary by company, by industry, and with the economic climate. Small firms and start-up companies have the most to gain by such arrangements, and the most to give to students in the way of broad perspective. Many graduate schools are surrounded by small companies started from university science and engineering programs.

John Armstrong, retired IBM vice president for science and technology, in "Rethinking the PhD," Issues in Science and Technology, Summer 1994.

---

## RESEARCH STRATEGIES AT NATIONAL LABORATORIES

Research in federal agencies is organized primarily to serve the scientific and technological objectives of their overall missions. Within that mandate, however, flexibility has evolved in recent years, especially among agencies whose missions have taken new directions. The evolution of missions is a natural consequence of broader societal change, such as the end of the Cold War and the growing urgency of environmental and energy issues.

National laboratories are maintained by many agencies, with the largest and best known funded by the Department of Energy (DOE), Department of Defense, National Institutes of Health, and National Aeronautics and Space Administration (NASA). Some of the facilities employ thousands of people and maintain the nation's most advanced technological equipment, affording unique opportunities for both research and training.

The national laboratories of DOE, the largest component of the national laboratory program, include those created to develop nuclear-weapons technology, beginning with the pioneering Manhattan Project. Many of the weapons laboratories have recently added multidisciplinary research programs in biology, medicine, chemistry, environmental science, energy efficiency, and other fields, diversifying the nation's research enterprise. For example, about half the research conducted at Lawrence Livermore National Laboratory (LLNL) and Los Alamos National Laboratory (LANL), originally focused entirely on nuclear-weapons research, is now unclassified. As an indication of their changed missions, LLNL and LANL were the first two laboratories to begin working on the Human Genome Project, in 1983. Another typical example is LLNL's new Center for Accelerator Mass Spectrometry, which is used by researchers in numerous nonprofit foundations, non-DOE agencies, and private firms for isotope-abundance measurements..

> IDR hasn't gone as well when we didn't have a team that was well integrated, when we still had a bunch of solo investigators without sufficient passion to solve the larger problem. Team members have to know that they bring only a portion of the answer and have to respect the contributions of all members. We can't have a physicist thinking "I do more important work" because they are using a supercomputer because a geologist is mapping rock formations with a colored pencil.
>
> Norman Burkhard, Lawrence Livermore National Laboratory

Although national laboratories engage in the same kinds of long-term fundamental research found in university settings, most of their work resembles the top-down, project- and budget-driven activities typical of industry.[12] As noted in one report, "the laboratories are . . . capable of forming large, interdisciplinary research teams needed for certain types of 'big science' problems even where large facilities are not involved. Universities are not generally as well equipped to assemble teams to conduct closely coordinated, interdisciplinary research over an extended period."[13]

Because many graduate students will eventually work on solving big problems with large teams, internships and other work experiences in government laboratories can add valuable career experience. Roughly 26,000

---

[12]Frosch, ibid. p. 419.

[13]Department of Energy. "Science and Engineering Roles" Chapter in Alternative Futures for the Department of Energy National Laboratories (known as the Galvin report) prepared for its chair by the Task Force on Alternative Futures for the Department of Energy National Laboratories. February 1995. *http://www.lbl.gov/LBL-PID/Galvin-Report/GalvinReport6. html#RTFToC50*.

scientists and engineers work in the 15 largest government research laboratories owned by DOE, for example. Some of the national laboratories are linked to or managed by research universities, so the national laboratory setting provides important opportunities for academic researchers to be involved in IDR. Some national laboratories have unique instrumentation for problem-solving, such as synchrotron facilities, and are engaged in solving problems of large magnitude and high risk, such as seeking novel sources of energy. Such large problems can be approached only by interdisciplinary teams that include special expertise.

### Some Models and Lessons from US National Laboratories

Although no sampling of national laboratories can truly represent the enormous breadth of activities at such facilities, many of them have the same ways of applying IDR to address complex problems, organizing their personnel and activities to facilitate IDR, and promoting practices of possible value to universities that wish to incorporate more IDR. The following discussion of these practices is distilled from the comments of leading scientists at three institutions:[14]

* Oak Ridge National Laboratory (ORNL), a DOE laboratory in Oak Ridge, Tennessee, was created in 1943 to produce plutonium for the Manhattan Project. It is administered by a limited-liability partnership of the University of Tennessee and Battelle. While continuing its weapons research, it now has multiple missions in materials, instrumentation, advanced computing applications, robotics, energy-technology development, computational biology, nanotechnology, environmental change, geographic information systems, and other fields.
* LLNL, in Livermore, California, was founded in 1952 as the nation's second nuclear-weapons laboratory (after LANL). Also funded by DOE and run by the University of California, it has a diverse portfolio of science and engineering programs.
* Jet Propulsion Laboratory (JPL), funded by NASA and managed by the California Institute of Technology, was founded in 1944 in response to Germany's V-2 program to develop rockets for the Allied war effort. It became part of NASA in 1958 and now manages the Mars Rover mission, Cassini Saturn mission, and other efforts to explore the Solar System and Earth.

---

[14]Thomas Wilbanks, corporate fellow, Oak Ridge National Laboratory; Edward Stone, former director, Jet Propulsion Laboratory; and Norman Burkhard, acting associate director, Energy and Environment Science Directorate, LLNL.

## Importance of IDR at National Laboratories

IDR has been important to all national laboratories since their foundation. They all use large, multidisciplinary teams to attack problems that require a wide array of skills, often in both science and engineering, and that are too complex for research teams based in any single discipline.

Former ORNL Director Alvin Weinberg compared the role of the national laboratories with research in other sectors as follows: Universities set their research priorities by the perspectives of academic disciplines; industrial organizations set R&D priorities according to marketing and profitability goals; and national laboratories set their priorities according to global, national, and social needs. These needs must often be addressed by R&D that is both multidisciplinary and too long term or risky to produce near-term results or profits.

## Strategies of National Laboratories in Recruiting and Organizing IDR Teams

Because of the interdisciplinary nature of their work, national laboratories tend to hire people who want to work on teams. As one manager said, "A lone investigator working on a single problem might have to turn out award-winning results to get the same pay and performance recognition as a team person." In hiring, the laboratories look first for people who are technically skilled; beyond that, they look for communication skills, writing skills, and evidence that they work well with people outside their own disciplinary space. Those who are hired but find that they do not want to work on teams usually "self-select" to move elsewhere.

Work at the national laboratories is often organized as a matrix system, with staff assigned to broad fields of science rather than single disciplines. Research programs are organized and promoted by cross-cutting program offices. Program leaders may set about addressing problems or topics by building teams from scratch. That is done by approaching people who have relevant skills and inviting them to discuss the problem at hand. Those who exhibit a passion for the problem and see clearly how their own work fits into a common vision tend to self-select for collaboration. The discussion groups may expand into local or multi-institutional IDR centers of excellence, often adding expertise from additional fields.

To facilitate IDR, JPL employs interdisciplinary scientists who are focused on broader scientific questions. The researchers function as "gluons" among the science teams, providing a broader view of science and systemwide issues.

IDR is becoming more important as we try to understand how systems work. While many fundamental, single-discipline questions remain to be addressed, science and engineering are ready to address much bigger questions, such as ecologic and planetary systems. No single discipline has the capability to even start addressing whole systems.

Edward Stone, Jet Propulsion Laboratory

## When IDR Works Well

As implied above, IDR works best when it responds to a problem or process that exceeds the reach of any single discipline or investigator. For example, astrobiology, a major NASA initiative to explore the origins and distribution of life, is a subject that requires the participation of multiple disciplines (see Box 6-2).

At LLNL, an urgent topic is the effect of global climate change on regional water supplies. Estimating such an effect requires diverse experts who can collaborate on a chain of linked questions: atmospheric scientists to set up global-climate models, computer experts to run the models, statisticians to do output analyses of precipitation, surface hydrologists to study river flow, groundwater hydrologists to study subsurface movement, aerosol physicists to study cloud structure, and so on. "We couldn't begin to address this topic without interdisciplinary collaboration," said Norm Burkhard, the project manager, "and even when we need specialists to bore down deep in a specific problem, they are usually successful only if they can talk about their work with the people around them."

## When IDR Is Less Successful

The commonest cause of underperformance of IDR is the failure of a team to gel or function collaboratively. That may happen for various reasons: individual members may place the importance of their own work ahead of the team vision, devalue the contributions of other team members, or lack leadership. Other contributing causes of lower-than-expected outcomes may be inadequate recognition for contributions to teams, low participation or understanding by senior staff members, inadequate time for participants to establish close working relationships, and insufficient funding.

On occasion, a culture gap between participating fields is not bridged. In the case of some early robotics research, for example, mechanical engineers and software engineers had widely different approaches. To the first group, a robot with adequate sensors had little need for software; to the second group, an abundance of mechanical sensors was a sign of inadequate software. Such cultural gaps must be bridged through persistent interaction and mutual efforts to understand other disciplines.

## How IDR Has Changed Over the Years

Answering research questions at national laboratories requires more disciplines and collaboration than in the past. The same is true at universities, where more individual researchers are working together on small teams. The researchers themselves are likely to have transcended disciplinary boundaries in their own work. "Thirty years ago, the difference between a physicist and a chemist was obvious," said Norm Burkhard. "Now we have chemists who are doing quantum-level, fundamental studies of material properties, just like solid-state physicists. There's almost no difference."

More research today is defined or driven by the priorities of funding. When funding is scarce, laboratories may respond by decreasing the size of projects and encouraging more "stove-piping" by disciplinary units—an unwillingness to branch out beyond their own confines. That reduces the ability of laboratories to support complex, expensive projects and to cross disciplinary boundaries.

## Lessons of National Laboratories for Academic Institutions That Wish to Facilitate IDR

Because so much interesting science of today involves complex systems, university researchers want to engage in the IDR required by systems questions. But national-laboratory scientists agree that IDR must be a valued part of institutional culture if it is to succeed. If a department or institution rewards only work that produces publications for journals in a narrow disciplinary field, academic researchers will respond accordingly.

One strategy that universities may adopt is to follow the practice of national-laboratory directors in setting aside funding to use as IDR seed money. At DOE laboratories, this seed money is important for launching projects in new directions. Universities could use such funding (which is now often used to hire new faculty) when existing faculty propose a major new initiative or interdisciplinary center. Without such startup assistance, it is difficult for established researchers to reorient their research, because funders may be hesitant to shift toward an unproven approach. In such cases, it is important for universities to lead, not follow, the funding agencies.

Another potentially valuable lesson is the use of sunset clauses. The National Science Foundation (NSF) Engineering Research and Science and Technology Centers have a 10-year life span, in recognition that they will support new subjects vigorously but not indefinitely (see Box 8-2). The University of California uses such a process for research centers and institutes that it runs: after 5 years, a panel of reviewers asks whether the program should remain an institute or it should begin a phaseout period with the objective of moving to a new subject.

Other steps suggested by the national-laboratory scientists are to

- Provide encouragement and rewards to move bright, early-career staff out of too-narrow disciplinary pursuits. For instance, an approach used in a few universities that run government laboratories is to put some tenure-track positions in issue-oriented "soft money" centers as a way to offer job security to promising nontenured staff with IDR interests.
- Encourage and reward team research rather than discouraging it. For instance, at least one division at ORNL has given every author of a joint publication the same performance credit as those who write single-author papers.
- In allocating discretionary research support, give priority to proposals that include and represent IDR.
- Encourage influential senior R&D staff to appreciate, participate in, and serve as role models for IDR, in part by making it an element in annual performance reviews.

Lessons have been learned from decades of hard experience about how to facilitate IDR. First, involve only people who find unraveling a complex transdisciplinary issue at least as important as their own discipline. Second, discourage "disciplinary entitlements," where something is accepted as truth because one discipline says so. Third, be sure all team members know that their reputations will be affected by the success or failure of the enterprise—that everybody's name will be on the product. Fourth, spend a lot of time in replacing disciplinary stereotypes with personal relationships and recognize the critical importance of leadership in both style and substance.

Thomas Wilbanks, Oak Ridge National Laboratory

## INTERDISCIPLINARY RESEARCH IN JAPAN

Japan's Ministry of Economy, Trade, and Industry (METI)[15] places heavy emphasis on IDR. Specifically, the National Institute for Advanced Interdisciplinary Research (NAIR) is one of 15 research institutions of the Agency of Industrial Science and Technology (AIST). The AIST laboratories concentrate on R&D programs judged to be capable of raising the level of Japan's technology.

NAIR was founded in January 1993 with an objective of pursuing IDR themes covering fundamental and frontier subjects of industrial science. It is portrayed as an innovative attempt to overcome institutional boundaries by bringing together scientists of diverse specialties—not only from research institutes under AIST and the Science and Technology Agency but also from universities and research organizations in the private sector.

Recent NAIR research projects include

- The Atom Technology Project (nanotechnology).
- The Cluster Science Project (experimental and computational study of the character of clusters).
- The Bionic Design Project (cell and tissue engineering and molecular machines).
- Next Generation Optoelectronics (large-capacity optical memory).

Each of these projects brings numerous disciplines together to solve specific cutting-edge problems of current interest.

NAIR management is based on four principles: extensive openness, flexibility and mobility of staffing, international collaboration, and objective evaluation of research progress. Although NAIR does employ researchers, most research staff members are drawn on a temporary basis from government, industrial, academic, and foreign organizations. That provides an interesting contrast with the US national laboratories, which support large permanent staffs.

## GOVERNMENT-UNIVERSITY-INDUSTRY RESEARCH COLLABORATIONS

As more faculty researchers become interested in applications of their research results and industries place greater emphasis on short-term outputs, new IDR partnerships are emerging between academe, industry, and government.[16] In general, the collaborations yield substantial benefits for

---

[15]The giant Ministry of International Trade and Industry, which had supported S&T research since its formation in 1949, lost power after the liberalization of trade and was reorganized as METI in 2001.

[16]Government-University-Industry Research Roundtable (GUIRR), "Overcoming Barriers to Collaborative Research: Report of a Workshop," Washington, D.C.: National Academy Press, 1999. University-government collaborations, such as the NSF-funded engineering research centers and science and technology centers, have generally succeeded in blending the two cultures. The growth of new government-university partnerships, however, has not been as rapid as the growth of industry-university partnerships.

all partners.[17] Box 3-3 provides an illustration of this for hard-disk-drive research.

University-industry collaboration, in particular, has proliferated over the last 2 decades, propelled partly by the Patent and Trademark Laws Amendments of 1980 and revisions, commonly referred to as the Bayh-Dole Act. One effect of these changes was to rationalize and simplify federal policy on patenting and licensing by universities of the results of publicly funded research.[18] A second contributing factor has been the revolutionary advances in university-based life-science research. Locating corporate research laboratories near major research universities creates more opportunities for these partnerships. As noted above, much of modern life science is inherently interdisciplinary, so these collaborations call for new, effective IDR strategies.

While the value of IDR partnerships is clear, practices for effective collaboration between universities and industry must be considered up front, including[19]

- Allocation of intellectual-property rights.
- Concerns over publication, copyright, and confidentiality.
- Regulation, liability, and tax-law issues.
- Concerns over foreign access.
- The involvement and best interests of graduate students.
- Infrastructure-related impediments to interdisciplinary and interdepartmental research.

Structuring and managing partnerships that produce gains for all partners take experience, careful planning, and continuing attention if universities, in particular, are not to risk compromising their educational focus.[20] Effective practices for surmounting such barriers include building trust between partners, efforts to understand the culture of the partner organization, attention to the misuse of students as "employees" of research sponsors, fair sharing of indirect costs, disposition of intellectual-property and patent rights to encourage the widest possible use of research tools, and

---

[17]Roessner, J. D. "University-industry collaborations: Choose the right metric," *Science's Next Wave*, June 1996.

[18]Mowery, D. C. "Collaborative R&D: How effective is it?," *Issues in Science and Technology*, Fall 1998. U.S. General Accounting Office, *Technology Transfer: Administration of the Bayh-Dole Act by Research Universities*, GAO/RCED-98-126, Washington, D.C., 1998.

[19]GUIRR, ibid. p. 7.

[20]For an extended discussion of this issue, see Bok, D. *Universities in the Marketplace: The Commercialization of Higher Education*, Princeton: Princeton University Press, 2003.

INNOVATIVE PRACTICE

**BOX 3-3   Establishing an Interdisciplinary Environment for Hard-Disk-Drive Research**

The best example of a product of industrial IDR is perhaps the hard disk drive (HDD) found in most computers and now beginning to appear in consumer applications and cell phones.

The first HDD was developed by IBM in the middle 1950s. It consisted of a spinning disk coated with a layer of small magnetic particles. An electromagnetic transducer positioned over the disk on an air bearing provided the writing field and inductive readout capability. The HDDs of today have the same basic design, but the medium is a thin magnetic metallic film, and readback is accomplished with a thin-film sensor whose resistance reflects the magnetic data pattern on the disk. The critical dimensions—the head-to-disk spacing, the thickness of the recording layer, and the spacing of data on the disk—are all in the range of nanometers, so it has become necessary for advances in one of these aspects to involve all the others. That requires the cooperation of materials scientists, mechanical engineers, chemical engineers, signal-processing engineers, and magnetism specialists.

There were many HDD companies in the 1980s. Many bought disks and heads and simply assembled the HDDs. Today, it is important to be vertically integrated on the basis of interdisciplinary technology development. By being vertically integrated, one can ensure that the heads and magnetic media are appropriately matched or perhaps compensated for by the design of the detection scheme.

To support such interdisciplinary technology, the industry has taken several steps. One is educational. In the early 1980s, it became obvious that traditional disciplines were not broad enough to train a "disk-drive engineer," Consequently, the industry encouraged and financially supported the formation of interdisciplinary centers in data storage. The most notable are at Carnegie Mellon and the University of California, San Diego. The centers bring together faculty that represent all the disciplines required in the design of high-performance storage systems. Curricula have been developed to expose students to all the scientific fundamentals required to produce this remarkable electromechanical device.

Because even the largest companies in the industry do not have expertise in all the disciplines required, the industry has pooled its resources through a consortium, the International Storage Industry Consortium, to develop technology road maps that identify where research is required to maintain the growth of the technology. The research is carried out by companies and universities that have the appropriate expertise. Thus, industry has, in effect, established a worldwide research environment to accomplish its interdisciplinary goals.

sensible agreements on publication delays to maintain the openness of the university research environment.[21]

## CONCLUSIONS

As suggested earlier, more contemporary data are needed to understand how IDR is managed in industrial and national laboratories. The prevalence of IDR has increased enormously since early studies on IDR in these settings was done, but yet even such fundamental questions as the following are not easily answered:

- How important to the success of IDR are the size and complexity of the organization?
- Does IDR work as well in small companies as in large ones?
- Does IDR work better for some types of problems than others?

In the absence of rigorous scholarly attention to such questions, we can still conclude that each sector performing and supporting IDR—academe, industry, and government—can learn from the best practices of other sectors. Researchers and administrators in institutions where IDR is unusual or neglected may be able to find helpful models in institutions where IDR is the norm, especially industrial and national laboratories. For example, they can observe how people behave when they are put together with others in teams, how researchers communicate across the barriers of knowledge domains, how large projects can be created and managed, and how projects can be disbanded when their usefulness comes to an end. They may also make wider use of other successful practices, for example, to

- Explore flexible organizational structures that permit shifting of resources and personnel to research subjects of highest promise.
- Establish reward systems that recognize outstanding performance in interdisciplinary research.
- Clarify and focus the mission of the laboratory or institution.
- Provide flexibility and support to small groups in seeking new knowledge.
- Organize laboratories not by discipline but by broader subjects of science or particular challenges.
- Use facilities and experts not available in their own institutions to solve specific problems.

---

[21]GUIRR, ibid. pp. 8-13.

In general, academe might find industrial practices that facilitate cognitive and social aspects of IDR and large-scale management of IDR helpful. One must recognize that the performance of IDR in academe occurs in its own particular institutional setting with its own conditions of rewards, budgeting, and especially responsibility for training the next generation of researchers.

## FINDINGS

Although research management in industrial and government settings tends to be more "top-down" than it is in academe, universities may benefit by incorporating many IDR strategies used by industrial and national laboratories, which have long experience in supporting IDR.

Collaborative interdisciplinary research partnerships among universities, industry, and government have increased and diversified rapidly. Although such partnerships still face substantial barriers, well-documented studies provide strong evidence of both their research benefits and their effectiveness in bringing diverse cultures together.

# 4

# The Academic Researcher and Interdisciplinary Research

I ndividual researchers involved in interdisciplinary research (IDR) re-quire a supportive environment that permits them to work in multiple disciplines and departments and to be fairly evaluated and rewarded for both their interdisciplinary and their disciplinary work. They have a re-sponsibility to explain and demonstrate the benefits of IDR, venture into new fields, and be open to the cultures and values of other disciplines.

The following sections condense numerous interviews, workshop dis-cussions, survey results, and firsthand experiences of committee members to portray in some depth the experiences of interdisciplinary students, postdoctoral fellows, and faculty members in academic institutions. Much of this material is based on anecdotal evidence, complemented by a large number of case studies and other reports in the literature, but its origin in experience can be instrumental in understanding the importance of provid-ing IDR-friendly environments at every stage of a scientific career.

Researchers need opportunities to train in two or more disciplines and to work closely with faculty members and students in each. Such cultural and intellectual immersion is a prerequisite to high-quality interdisciplinary work. Researchers may need to spend considerable time on activities (teach-ing, research, committees, and community service) outside their home department. In the committee's survey of those interested in IDR, over half indicated that after training in a specific field they had sought training in additional fields through either postdoctoral fellowships, further advanced degrees, or day-to-day interactions in interdisciplinary projects. People whose home departments do not recognize, encourage, and reward such activities may not be willing to make the extra effort required for interdisciplinary activities.

---

**Convocation Quote**

The most interesting observation is that the students are the integrating glue. Graduate students, undergraduates, and postdocs are the ones that go between the laboratories that make things happen.

Harvey Cohen, professor of pediatrics, Stanford School of Medicine and chair, Interdisciplinary Initiatives Program

---

## UNDERGRADUATES

Undergraduates can have a rich educational experience when they learn about and in more than one discipline, especially when education is complemented by research experience. Students at Brown University have shown a consistent interest in interdisciplinary programs (Figure 4-1). At Columbia University the number of students majoring in interdepartmental or interdisciplinary programs has increased dramatically over the last 10 years (Figure 4-2). Harvard University students are also increasingly interested in interdisciplinary studies: the number of undergraduate joint concentrations in chemistry and physics has risen from 14 to 45 over the last 15 years (see Box 9-2). At Stanford University a multiyear decline in the number of students majoring in earth science was reversed when the major, originally based in the single discipline of geology, was reformulated into the interdisciplinary program "earth systems" (see Figure 8-1).

University policies can facilitate or hinder students' ability to learn about IDR and to take double majors, take courses in other schools, or custom-design their majors and participate in IDR. For undergraduates to gain deep interdisciplinary insights, they need to work with faculty members who offer expertise both in their home disciplines and in the interdisciplinary process (see Box 4-1). In the committee's survey, the top recommendations to students were to cross boundaries between disciplines (25 percent), to take a broad range of courses (23.4 percent), but also to develop a solid background in one discipline (12.3 percent). Respondents overwhelmingly recommended that educators incorporate interdisciplinary concepts in course curricula (Figure 4-3). But structural roadblocks can impede faculty in offering the team teaching and co-mentoring that are essential to undergraduate education. Another barrier in some disciplines,

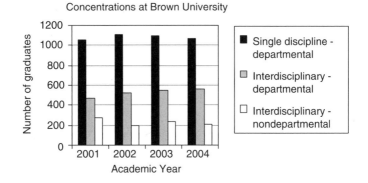

FIGURE 4-1 Consistent undergraduate interest in interdisciplinary studies at Brown University.

NOTES: In consultation with an appropriate faculty member, students at Brown University devise a concentration program centered on a discipline or disciplines, problem or theme, or broad question; they may also select a standard departmental concentration. Interdepartmental concentrations make up about one-third of the standard programs. Students may also design their own concentration, in which case a written proposal presenting a statement of the major objectives of the concentration program and a list of the specific courses to be taken are signed jointly by the student and faculty adviser and submitted to the College Curriculum Council for approval. Standard concentration programs require only the approval of the appropriate department or committee. In this environment, consistently over 40 percent of students graduate with an interdisciplinary concentration, 30 percent from departmental and 10 percent from non-departmental programs.

SOURCES: Data provided by the Office of the Dean of the College, Brown University, June, 2004, Brown University Undergraduate Concentration Requirement: General Information *http://www.brown.edu/Administration/Registrar/concentration. html*; Brown University Dean of the College, Academic Advising and Support, Concentration Programs *http://www.brown.edu/Administration/Dean_of_the_ College/DOC/s1_advising_support/conc_codes.shtml*.

such as engineering, is a curriculum so packed with required courses that it is difficult to take electives or concentrations in disciplines outside the major.

## GRADUATE STUDENTS

Many researchers begin serious involvement in IDR as graduate students. They may obtain a master's degree in a second subject; for example,

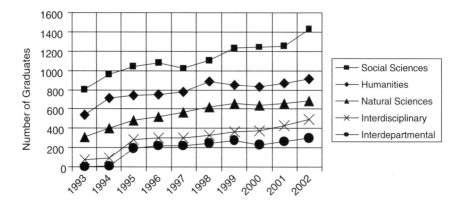

FIGURE 4-2    Trends in undergraduate interest in interdisciplinary studies at Columbia University.
NOTES: There has been a marked increase in the number of undergraduates at Columbia graduating with interdisciplinary or interdepartmental majors or concentrations. This increased student interest followed university administration promotion of and faculty interest in interdisciplinary research and teaching in the 1990s. The 9.7 percent average annual increase in interdisciplinary program majors and concentrations has outpaced interdepartmental (6.7 percent), and departmental majors and concentrations (4.8 percent).
SOURCE: Data provided by the Office of the Vice President for the Arts and Sciences, Columbia University, May 2004; average annual increase was calculated for the years 1995-2002 and does not include the very large increase in majors and concentrates in interdisciplinary and interdepartmental programs that occurred between 1993-1995.

economics or psychology majors may take an MS in statistics to deepen their understanding of statistical analysis. Such involvement depends on finding multiple advisers who are interested in working together and with the students; on gaining sufficient training in the "other" discipline, which calls for the support of the home department. For doctoral students working in an IDR team environment, fulfilling the requirements for a PhD qualifying examination or dissertation in the home department may require extra planning and coordination between departments (see Box 4-2). There may be barriers to entry, such as admissions policies, that are biased against students whose undergraduate degree is not in the same discipline as the proposed graduate degree.

TOOLKIT

## BOX 4-1 IDR Immersion Experiences: Summer Research Opportunities

One of the common themes that runs through any discussion of interdisciplinary interactions is the learning of new disciplinary languages and cultures. One way to accomplish that is to immerse oneself in a new discipline. There are several examples of immersion research experiences; most tend to be summer internships. All feature an infrastructure and an informal, interactive scientific community that allows researchers to launch into research almost immediately upon their arrival and to develop long-lasting research networks and collaborations.

The Berkeley Mathematical Sciences Research Institute (MSRI)[a] offers postdoctoral scholarships in conjunction with Hewlett Packard and Microsoft Research. MSRI postdoctoral fellows are in residence for 5 months. Microsoft Research has an inhouse internship program[b] for research on human-computer interactions.

Woods Hole Marine Biological Laboratory offers a visitors program[c] in which summer researchers—graduate students and postdoctoral scholars and professors—enjoy 3 months of research without academic responsibilities. In 2003, 139 principal investigators and 201 other researchers from 144 institutions in 18 countries converged on MBL to perform research in marine biology, neuroscience, and ecosystems.

The Shingobee Headwaters Aquatic Ecosystems Project (SHAEP)[d] in Minnesota offers summer interdisciplinary immersion experiences for researchers interested in hydrology. Developed in 1987 around instrumentation installed and funded by US Geological Survey researchers, SHAEP is based on the concept that proper management of water resources requires knowledge about atmospheric water, surface water, groundwater, and how these resources function as an integrated system. There are no dedicated faculty members, but a full-time staff coordinator was only recently hired. People using the site share equipment but must bring their own funding. There are no constraints on the number of people participating or on their disciplines. SHAEP has instrumented similar interdisciplinary sites in Nebraska, North Dakota, and New Hampshire.[e]

Yet another summer interdisciplinary immersion experience can be had at the University of Michigan's Biological Research Station (UMBS). At any given time, there are usually 250 people present, a mix of resident researchers and short-term researchers. The site was developed by biologists, but atmospheric scientists recognized that the instrumentation available was also useful for their research. Weekly talks were initiated, and the two groups forged an understanding on research terminology, methodology, and the measurements that each group was capable of taking. The talks inspired the Biosphere Atmosphere Research and Training (BART) and Integrative Graduate Education and Research Traineeship (IGERT) program, a multidisciplinary doctoral training program.[f] During two summers at UMBS, BART students from over 15 universities participate in educational research programs at the biosphere-atmosphere interface.

---

[a] http://www.msri.org/.
[b] http://research.microsoft.com/aboutmsr/jobs/internships/.
[c] http://www.mbl.edu/research/summer/index.html.
[d] http://wwwbrr.cr.usgs.gov/projects/SHAEP/index.html.
[e] http://www.npwrc.usgs.gov/clsa/index.htm.
[f] http://www.bart-wmich.org/.

SURVEY

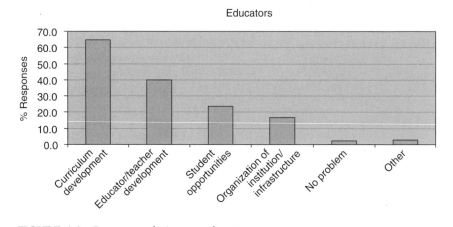

FIGURE 4-3    Recommendations to educators.
NOTES: Survey Question: If you could recommend one action that educators could
take that would best facilitate interdisciplinary research, what action would that
be? Survey respondents (n = 190) recommended that educators develop curricula
that incorporate interdisciplinary concepts (64.7 percent), take part in teacher-
development courses on interdisciplinary topics (40 percent), and provide student
opportunities in IDR (23.7 percent). These recommendations echo other recent
reports and statements.
SOURCES: Gregorian, V. 2004, "Colleges must Reconstruct the Unity of Knowledge" The Chronicle Review, Vol. 50/39, p. B12; Kellogg Commission, "Renewing
the Covenant: Learning, Discovery, and Engagement in a New Age and Different
World," March 2000, *www.nasulgc.org/publications/Kellogg/Kellogg 2000-
covenant.pdf*; Bartlect, T. 2004, "What's Wrong with Harvard?" *The Chronicle of
Higher Education*, Vol. 50/35. p. A14.


The last step may be hindered when examiners view a student's work
from the viewpoint of only a single discipline. One study concludes that
although graduate students report that interdisciplinary activities have adverse effects on their careers, they are convinced of the value of IDR; the
study also describes graduate students and postdoctoral scholars as "essential links" in the skill networks of IDR centers.[1]

---

[1]Rhoten, D. Final Report, National Science Foundation BCS-0129573: A Multi-Method
Analysis of the Social and Technical Conditions for Interdisciplinary Collaboration. September 29, 2003. Available at: *http://www.hybridvigor.net/interdis/pubs/hv_pub_interdis-2003.
09.29.pdf*.

INNOVATIVE PRACTICE

## BOX 4-2   Interdisciplinary Departments Train Interdisciplinary Students

The School of Life Sciences (SOLS) at Arizona State University (ASU) has taken a directed approach to changing the culture of the unit and in the process has affected how graduate students are being educated. Within SOLS, discipline-based and interdisciplinary researchers are developing a culture that supports both disciplinary and interdisciplinary approaches to research.

The school has 80-85 faculty members who are organized into six faculties that have few fixed boundaries. The six faculties have no budget lines, and each year members are allowed to move freely among them. Among the faculty members are historians of science, bioethicists, policymakers, and philosophers of science. Faculty members in the humanities and social sciences are imbedded in the department, and this allows a truly interdisciplinary educational experience that includes such concentrations as "Biology and Society" and such research groups as "Human Dimensions of Biology."

Students in ASU's urban-ecology IGERT write one chapter of their dissertation jointly with a student who is also in the program but in another department. Coauthorship, one obstacle to IDR, would be easier to overcome if researchers were involved in such collaborations during their training. Students conducting IDR also benefit greatly from the guidance of mentors in the several disciplines represented. Comentoring allows students to have direct relationships with researchers in the different fields while synthesizing the training and advice to form their own skills and experiences for their future IDR goals.

[a]James Collins. Convocation on Facilitating Interdisciplinary Research. Washington, DC, January 30, 2004.

## POSTDOCTORAL SCHOLARS

A postdoctoral experience often provides the best opportunity for researchers to train deeply in a new discipline. The training provided in postdoctoral years can provide skills and knowledge beyond those acquired by graduate students, which are focused on the home discipline. Respondents to the committee's survey encouraged postdoctoral scholars to broaden their skills and knowledge base (see Figure 4-4). Despite committee interviews that indicate heightened interest in IDR among postdoctoral scholars, progress toward interdisciplinary expertise may be slowed by a relative shortage of interdisciplinary postdoctoral fellowships. Moreover, a potential fellow may not be sufficiently knowledgeable about the secondary discipline to be useful to a potential mentor.

## SURVEY

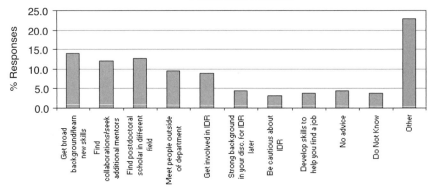

FIGURE 4-4   Recommendations for postdoctoral scholars.
NOTES: Survey Question: If you could recommend one action that postdoctoral scholars could take that would best facilitate interdisciplinary research, what action would that be? Respondents (n = 157) encouraged postdoctoral scholars to get a broad background and learn new skills (14.0 percent), to find postdoctoral fellowships in fields different from their own graduate work (12.7 percent), and to develop collaborations and seek additional mentors (12.1 percent). A recent report listed similar recommendations.
SOURCE: NRC, "Enhancing the Postdoctoral Experience for Scientists and Engineers," National Academy Press, 2000.

A productive transition from graduate student in one department to postdoctoral fellow in another frequently requires "translators" who can provide training in the postdoctoral environment. Such trainers need to be well enough versed in both disciplines (and in their methodological and knowledge differences) to simultaneously help the postdoctoral scholars obtain new skills and knowledge and share perspective about research issues (see Boxes 4-3 and 4-4).

A special challenge for all postdoctoral researchers, whether disciplinary or interdisciplinary, is to produce an expected number of publications and other indicators of productivity. The additional training in a new field needed for an interdisciplinary researcher may reduce a postdoctoral scholar's apparent productivity relative to that of a scholar who focuses on a single discipline. As a result of the lower productivity, they may require more time and assistance in finding faculty positions after the postdoctoral

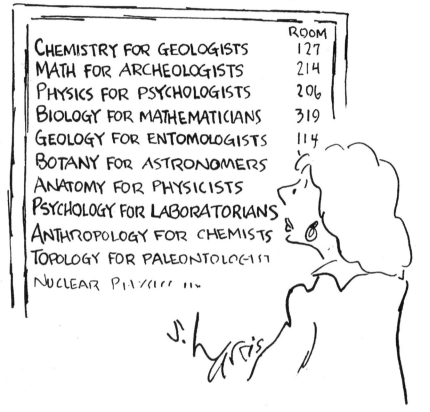

period, although postdoctoral scholars interested in pursuing nonacademic careers may find their employability enhanced by their interdisciplinary experiences.

## HIRING

Candidates for tenure-track positions who are interested in IDR face the additional challenge of finding departments that feel that the candidates "belong" with them. Universities vary in their willingness to offer joint positions; for example, the mathematics department at Stanford University does, but the physics department does not. When asked whether their institution made joint appointments in which salary support was shared between hiring units, 58 percent of respondents to the committee's individuals survey and 63 percent of respondents to the provosts' survey said yes. More than 75 percent said that 0-10 percent of faculty members held joint appointments.

Although joint appointments may be the only recourse for some, these researchers may find themselves serving two masters and satisfying neither. For example, universities may become concerned when a faculty candidate's annual research productivity has been lower than that expected of single-discipline candidates. Some universities and research supervisors offer mentoring and active assistance in helping departments to assess the value and substance of the work of IDR candidates.

---

**INNOVATIVE PRACTICE**

**BOX 4-3   The Global Environmental Assessment Project**

The Global Environmental Assessment (GEA) Project,[a] based at Harvard's Kennedy School of Government, ran an interdisciplinary research and training program funded by a 5-year grant from the National Science Foundation. To help to build a next generation of professionals trained in and sensitive to the unique problems of linking science and policy on global environmental problems, the GEA Project recruited fellows through an international competition open to natural scientists, social scientists, and professional-school students. A unique aspect of the program was the commitment to generational change in the visibility and acceptability of IDR, by bringing together a "critical mass" of young scholars with the intent to foster interdisciplinary and international collaborations during this formative stage in their careers.

Fellows were exposed to interdisciplinary methodological and professional approaches and perspectives in the year-long training experience, which consisted of discussions of key papers from various intellectual perspectives, presentations of research by GEA faculty and visitors, and an introduction to the science and policy of the specific issues studied each year. During the first 2 months of fellows' residence, seminars introduced them to program faculty and provided an early opportunity to discover how different are the things "taken for granted" in conceptualizing and pursuing research. In November and December, fellows designed their research projects, which they then pursued in field research through March. Results were used as input to annual GEA Project workshops at which practitioners, users, and scholars of environmental assessment engaged in off-the-record discussion comparing insights and experiences. Fellows' papers were

**Convocation Quote**

Our students are marketable as civil engineers, hydrologists, forest ecosystem biologists, or fishery scientists. But they are very subversive in that they are trained in a very different way. So, you get a hydrologist, who knows not only what he or she is supposed to know in a civil engineering department, but a hydrologist who can deal with the climate dimension and who can connect it to the societal dimension.

Ed Miles, professor of Marine Studies and Public Affairs,
University of Washington

posted on the GEA Project's public Web site, and the best of them have been revised for inclusion in the three volumes of final output from the project.[b]

The project graduated 37 fellows in 12 disciplines (anthropology, business and management, economics, engineering, environmental sciences, geography, law, oceanography, physics, political science and government, public policy, and science and technology studies) and of 10 nationalities (American, Australian, British, Bulgarian, Canadian, Danish, Dutch, German, Indian, and Swiss). There were 15 predoctoral fellows, 16 postdoctoral fellows, one practitioner, and five faculty fellows. All were readily able to get employment, and most US fellows have academic jobs (see table below). A network of alumni fellows is maintained to encourage continuing collaboration.

Current Jobs held by 37 Former GEA Project Fellows

|  | US | International | Total |
|---|---|---|---|
| Academe | 19 (51%) | 5 (14%) | 24 (65%) |
| Research Institutes | 1 (3%) | 7 (19%) | 8 (22%) |
| Government | 2 (5%) | 1 (3%) | 3 (8%) |
| Unknown | 0 (0%) | 2 (5%) | 2 (5%) |
| Total | 22 (59%) | 15 (41%) | 37 (100%) |

[a]The GEA Project Web page is *http://www.ksg.harvard.edu/gea.*
[b]Mitchell, R. B., Clark, W. C., Cash, D. W., and Alcock, F., eds. Forthcoming. *Global Environmental Assessments: Information, Institutions, and Influence.* Cambridge: MIT Press. Jasanoff, S., and Martello, M. L., eds. 2004. *Earthly Politics: Local and Global in Environmental Governance.* Cambridge: MIT Press. Farrell, A., and Jäger, J. eds. Forthcoming. *The Design of Environmental Assessments: Choices for Effective Processes.* Washington, D.C.: Resources for the Future.

INNOVATIVE PRACTICE

**BOX 4-4    The Institute for Mathematics and Its Applications**

The Institute for Mathematics and Its Applications (IMA) at the University of Minnesota was founded in 1982 with a grant from the National Science Foundation.[a] Its primary mission is to increase the impact of mathematics by fostering IDR. IMA's postdoctoral program was created to provide opportunities for mathematical scientists near the beginning of their careers who have a background or an interest in research involving applications of mathematics (e.g., mathematics of materials, genomics, networks, and financial engineering). IMA postdoctoral fellowships run 1-2 years and provide an annual salary of $45,000 and a travel allowance. There have been 191 postdoctoral members since IMA was founded in 1990 and 35 additional scholars in the IMA industrial postdoctoral program.[b]

IMA focuses on two important factors to ensure that postdoctoral scholars have a positive experience. First, it creates a focused scientific atmosphere built around a yearly thematic program in which one broad field of quantitative interdisciplinary science is studied; this offers a unique environment for the postdoctoral scholar to become truly immersed in a problem or question. Second, it dedicates a great deal of time to the mentoring of its postdoctoral scholars, involving both long-term visitors and local faculty members in mentoring roles. The industrial postdoctoral program offers recent mathematics PhDs the opportunity to work half-time in an industrial research laboratory while performing academic work. Postdoctoral scholars in this program receive mentoring from both industrial and academic coworkers.

In a continuing effort to evaluate, document, and improve its program, IMA collects followup information from its postdoctoral scholars in the form of surveys and requests for reports. For example, in spring 2003, IMA surveyed postdoctoral scholars from 2000-2001. On a scale of 1 to 5, 12 of the 13 responded with an average of 4.5 to the question, "Was your research more interdisciplinary because of the IMA?" They also agreed strongly with the statements "Interaction and collaboration were well facilitated by the IMA" and "I made useful contacts at the IMA."[c]

---

[a]IMA home page: *http://www.ima.umn.edu/*.

[b]Complete lists of the IMA postdoctoral members and IMA industrial postdoctoral scholars with current affiliations can be found at *http://www.ima.umn.edu/people/all-reg-postdocs. html* and *http://www.ima.umn.edu/people/all-ind-postdocs.html.*

[c]Douglas N. Arnold, Director, IMA. Personal communication, March 26, 2004.

## JUNIOR FACULTY

Many universities and departments appreciate the value of IDR but expect interdisciplinary faculty members to do "double duty": to first meet the usual obligations of disciplinary and departmental activity—including publications, teaching, and service—and then find additional time for IDR. Some junior faculty members achieve this by doing their interdisciplinary work in a disciplinary way, such as publishing mathematical results related to their field in a mathematics journal rather than in a journal in their field; others fortify their credentials by doing purely disciplinary research that is not related to their interdisciplinary interest.

Faculty members also feel pressure with regard to activities outside their departments. For example, interdisciplinary teaching, especially at the graduate level, often involves activities that are not recognized or rewarded by the home department, including

- Service and committee work.
- Teaching courses with other faculty members.
- Teaching courses in other departments.
- Teaching courses to attract and train doctoral students in the faculty members' own fields of research.

Such activities may be considered "extra" and earn little or no credit. In addition, faculty members might not be permitted to advise graduate students in other departments even if they would be the most appropriate mentors. Similarly, coadvising students, which is often the best way to train in IDR (see Box 4-5), can be difficult or discouraged by the institution. These pressures can affect student advisees and thus faculty members' own research productivity. The issues become more complex when a nontenured faculty member has a joint appointment and must seek tenure in two departments.

Involvement in IDR provides a number of benefits, including the opportunity to participate in unique projects and to build collaborative relationships with peer faculty members in other departments.

## GAINING TENURE

An interdisciplinary faculty member seeking tenure often faces two challenges beyond those faced by members working in a single discipline. Indeed, tenure and promotion criteria were listed as the top impediment to IDR by respondents to the committee survey (see Figure 4-5). First, as suggested above, IDR done by the candidate may not be valued sufficiently to compensate for a lower output of disciplinary research. Publications and other activities not recognized as being in the home department's discipline may be considered valuable but not sufficient for tenure. Similarly, reward systems at the level of dean and vice president of research do not necessarily reward IDR programs and activities.

Second, it can be difficult to find reviewers who understand the overall quality of the work, which usually lies outside the expertise of people on the tenure evaluation committee—that is, members of the department. In such

INNOVATIVE PRACTICE

**BOX 4-5   Combining Interdisciplinary Research
and Graduate Education[a]**

In 1993, the College of Agriculture at Pennsylvania State University (PSU) benchmarked its programs with six other life-sciences departments. The goal of this exercise, which took almost 2 years, was to determine and set priorities among needed improvements at PSU. The benchmarking committee proposed to the university that a life-sciences institute be created that was truly interdisciplinary. In 1996, PSU dedicated $5 million to the effort to hire new faculty members, to create an interdisciplinary graduate program, and to build shared technical resources. How was it funded? The provost charged all departments to come up with a 10% reduction in budget—the savings were recycled to the university—and then engaged the faculty in determining new initiatives.

The resulting Huck Institutes of the Life Sciences[b] is a virtual organization comprising seven of PSU's colleges. Like the Program on the Environment and the Center for the Neural Basis of Cognition (see Box 9-5), the Huck Institutes does not have faculty lines but instead collaborates with colleges and departments to cohire new faculty members. As at the University of Illinois at Urbana-Champaign Beckman Institute for Advanced Science and Technology (Box 5-6), core facilities staffed by PhD-level directors attract faculty members to work at the Huck. In the last 7 years, the Huck helped to hire 50 new faculty members, providing 50 percent of the starting salaries and 50 percent of the startup costs. The Huck form an agreement with a department as to how faculty members will participate with the Huck. Built into the agreement is a biannual evaluation by the department chair that focuses on undergraduate education, graduate education, and research inte-

cases, it is essential to include letters from outside reviewers who can adequately explain the importance of the work. Accepting such letters may require departments to change policies that limit external letters to those written by members of equivalent departments.

**Convocation Quote**

For tenure, yes, it is risky. You spend a lot of time doing all this groundwork. Here I am, picking the rocks from the field for two years. I finally get the pile and I finally get to plant the seeds. You have to be at an institution where that type of effort is respected and you also have to have enough projects that you know are going to succeed, that you can afford to risk some time on those that you are not so sure about.

Victoria Interrante, professor of computer science and engineering
at the University of Minnesota

gration. If at the 5-year mark a faculty member has not contributed to the Institutes, Huck will pull its funding regardless of the quality of his or her research.[c]

The Institutes encourage collaborative research alliances through a variety of mechanisms that integrate graduate training and research across disciplines. PSU provided up to $150,000 in seed money to faculty members to develop funding ideas and proposals in which faculty members from both physical and life sciences had to team up. That led to the development of nine core areas and graduate training programs in and across each.

The Huck Integrative Biosciences Graduate Degree Program offers students a venue to learn about and work in multiple disciplines. Meetings between people in different departments, in different colleges, and even on different campuses are supported by modern telecommunications facilities and equipment. Students in the program have to identify two advisers in at least two colleges or departments.[d] To promote interdisciplinarity and innovativeness in research and graduate education in the life sciences at PSU, the Huck Institutes is working with faculty members to develop interdisciplinary graduate groups; these may generate new options in existing graduate programs, new interdepartmental programs, or options in a new integrative-biosciences degree program of the Huck Institutes.

---

[a]Partially derived from staff interview with institute Director C. Channa Reddy, November 11, 2003.

[b]Huck Institutes of the Life Sciences home page *http://www.lsc.psu.edu/*.

[c]For more information on Huck Institutes organization, governance, structure, and budgeting and graduate education policies, see *http://www.lsc.psu.edu/proppol.html*.

[d]Pell, E. J., Reddy, C., and McGrath, R. T. (2004) Interdisciplinary Education and Research: Penn State's Huck Institutes of the Life Sciences. Poster presented at the Convocation on Interdisciplinary Research, Washington DC, January 29, 2004.

---

The contribution of the interdisciplinary researcher may also be questioned by a department in which collaborative work is not the norm. For example, in mathematics, single-author papers are the norm and are an important step toward tenure, whereas in chemistry coauthorship is the norm. In other fields, papers may have many senior and junior coauthors. The difficulty in parsing contributions may be mitigated when review panels understand how to "read" the various contributions of researchers in interdisciplinary collaborations, but this is complicated by cultural differences among fields. For example, in some fields (notably mathematics and computer science), the author order in publications is explicitly alphabetical; in other fields, the author order indicates the importance of the contributions. In many fields, the best work is customarily published in journals; in other fields, such as computer science, conferences are the most prestigious publishing outlets. Those cultural differences can complicate tenure review of researchers who focus on IDR.

SURVEY

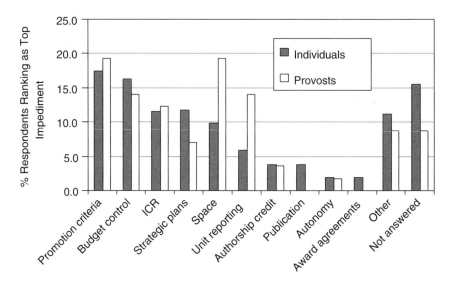

FIGURE 4-5    Top impediments to IDR.
NOTES: When asked whether there were impediments to IDR at their current institutions, 70.7 percent of the respondents answered yes, 23.2 percent answered no, and 6.2 percent did not know or did not answer. Respondents were provided a list (from Feller, 2002) and asked to rank the top five impediments to IDR at their institutions. Although the impediments cited are similar to the preconditions for IDR discussed by Klein and Porter (2002), it is interesting that "individuals" and provosts ranked impediments differently. Furthermore, impediments often mentioned in research literature–authorship credit and publication–were among the lowest ranked by both respondent groups. The impediments that were most often ranked first by "individuals" were promotion criteria, budget control, indirect cost returns (ICR), and compatibility with strategic plans. For provosts, the top impediments were promotion criteria, space allocation, budget control, and unit reporting. These differences reflect the perspectives of researchers looking for more control of their research interactions and provosts who are charged with having a global view of the university research portfolio.

Despite the apparent disadvantages, Rhoten et al. reported that interdisciplinary researchers spend about 50 percent of their total work time on extradepartmental activities related to IDR centers. About 30 percent of their sample reported that their interdisciplinary affiliations had not helped or had hindered their careers.[2]

---

[2]Rhoten, D. ibid. 2003.

## TENURED PROFESSORS

For professors who have secured tenure and would like to pursue IDR, a critical step is to immerse themselves in the "other" field so that their work can be of the best quality and have the greatest impact. That takes substantial time–not only in learning the language of other disciplines but also in learning new value systems, aesthetics, tastes, and methods. Establishing close relationships with researchers in another discipline on the other side is critical to the productivity and quality of a researcher's work. Finding appropriate collaborators can be difficult (but see Box 4-6), especially when they work at distant institutions. Time away from regular departmental activity is helpful for immersing oneself in another field and developing the kinds of collaborations that form the foundation of much IDR.

The top recommendations for principal investigators listed by survey respondents were to increase leadership and team-forming activities (44.1 percent) and to build networks with researchers in other disciplines (20.4 percent) (see Figure 4-6).

**SURVEY**

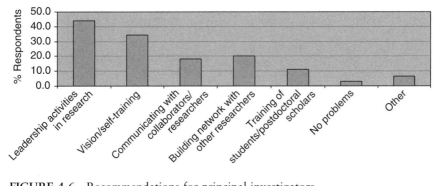

FIGURE 4-6  Recommendations for principal investigators.
NOTES: Survey Question: "If you could recommend one action that principal investigators could take that would best facilitate interdisciplinary research, what action would that be?" The top two recommendations for principal investigators given by survey respondents (n = 186) were to increase leadership and team-forming activities (44.1 percent) and to develop and clearly state their research goals and their overall vision (34.4 percent). Reports and evaluations of IDR programs have yielded similar suggestions. (See Boxes 2-4 and 8-2.) and Klein, J. T. and Porter, A. L. "Preconditions for Interdisciplinary Research." *In:* International Research Management Studies in Interdisciplinary Methods from Business, Government, and Academia, Eds. Birnbaum-More, P. H., Rossini, F. A., Baldwin, D. R. New York: Oxford University Press. 1990. pp. 11-19.

---

**TOOLKIT**

### BOX 4-6   Creating and Managing Interdisciplinary Collaboration

Several institutions have decided that rather than let random molecules collide, they would help students and researchers with similar interests collaborate. The Fred Hutchinson Cancer Research Center (FHCRC) established an Interdisciplinary Research and Training Initiative in 1996,[a] which includes a Dual Mentor Program for graduate students and postdoctoral scholars, a Joint Degree Program, and a Pilot Project Fund Program (which awards about eight to 10 grants of $20,000-25,000 each year). It also funds the Interdisciplinary Club.[b] Faculty members, overseen by staff associate Karen Peterson, handle the administrative aspects of these programs.[c,d]

The Dual Mentor Program is supported by a National Institutes of Health Training Grant (T32) written by several FHCRC faculty. It has just had a successful competitive renewal. The program has funded nine graduate students, four of whom have received their PhDs, and 10 postdoctoral fellows. There are also a few privately funded fellowship slots for which international postdoctoral scholars and graduate students are encouraged to apply.

The Joint Degree Program was started in 2000. By early 2004, five graduate students had received their MS in epidemiology and were working toward their PhD in molecular and cellular biology or microbiology. Peterson was not only involved in establishing this program but also organizes an epidemiology course that every potential joint-degree student attends before applying to the program.

FHCRC scientists have the opportunity to observe daily rounds for 2-4 weeks in the Observing Stem Cell Transplant Rounds Program. This program has inspired collaborations and new career directions among faculty members and fellows, and some of the participating postdoctoral scholars decided to pursue clinical or translational research.

---

Continued professional advancement is also sometimes harder for those doing IDR because recognition comes from established disciplines rather than from younger or unformed fields. Thus, promotion to full professor can be more difficult for interdisciplinary researchers than for disciplinary researchers for the same reasons that the tenure process is more difficult. There are also fewer honors and awards given by professional societies for IDR than for disciplinary research (but see Boxes 7-2 and 7-3). Finally, fellowship nominations even in multidisciplinary societies, such as the American Association for the Advancement of Science, are initiated in disciplinary committees, so interdisciplinary researchers often obtain fellowship status long after disciplinary researchers of comparable quality.

Another key barrier is the attitude of other senior faculty members

Peterson also develops courses and intrainstitutional symposia on such strategic topics as proteomics, biostatistics, cell signaling, immunology, and epigenetics and cancer. The goal of these courses is to help attendees to become better collaborators by learning the main concepts, approaches, and language of a field. The courses are taught by invited faculty members and postdoctoral scholars at FHCRC and other local institutions. Attendees include graduate students, postdoctoral scholars, and faculty members.

In January 2004, Peterson began facilitating faculty team development for large IDR grants. Her role is to identify calls for proposals that may be of interest to FHCRC faculty and to alert the center and division directors. She identifies faculty members who may be interested, invites them to participate, and then "gets out of the way" to let the faculty develop the proposals. Thus far, she has worked to identify faculty members for NIH programs in integrative cancer biology and in early detection and molecular imaging.

The Student-Postdoc Advisory Committee (SPAC) provides opportunities to promote interdisciplinary programs to FHCRC's approximately 100 graduate students and 300 postdoctoral scholars. SPAC also offers travel awards that give preference to attending interdisciplinary conferences and course scholarships that many awardees use to cross-train in fields, such as computer science and statistics.

---

[a]FHCRC Interdisciplinary Home Page *http://www.fhcrc.org/science/interdisciplnary/*.

[b]Paulson, T. (2003) Grassroots Interdisciplinary Training: The FHCRC Interdisciplinary Club. Science's Next Wave, Posted January 3, 2003 *http://nextwave.sciencemag.org/cgi/content/full/2002/12/30/7*.

[c]Peterson, K. (2004) The Interdisciplinary Research and Training Initiative at the Fred Hutchinson Cancer Research Center. Poster presented at the Convocation on Facilitating Interdisciplinary Research, Washington, DC. January 29, 2004.

[d]Karen Peterson, Personal Communication, April 23, 2004.

---

toward IDR. Some are openly scornful, claiming that it lacks the depth of discipline-centered research. This can be a serious barrier to junior and senior faculty members, as well as graduate students.

## CONCLUSIONS

This chapter is meant to convey a feel for the experience of the individual researcher that is not available in quantitative form. If, as the committee believes, the cumulative effect of the specific obstacles to IDR described here is larger than any single obstacle might suggest, understanding the character and source of each obstacle becomes a high priority for institutional leaders. The next chapter will examine those obstacles from the point of the view of the institution and suggest ways to adapt institutional structures to facilitate IDR.

## FINDINGS

Successful interdisciplinary researchers have found ways to integrate and synthesize disciplinary depth with breadth of interests, visions, and skills.

Students, especially undergraduates, are strongly attracted to interdisciplinary courses, especially those of societal relevance.

## RECOMMENDATIONS

### Students

**S-1: Undergraduate students should seek out interdisciplinary experiences, such as courses at the interfaces of traditional disciplines that address basic research problems, interdisciplinary courses that address societal problems, and research experiences that span more than one traditional discipline.**

For example, students can

*   Begin preparation for IDR through an IDR project or summer IDR experience.
*   Approach interdisciplinarity by first gaining a solid foundation in one discipline and then adding disciplines as needed. Additional courses provide opportunities to understand the culture of other disciplines, gain new skills and techniques, and network with other researchers.

**S-2: Graduate students should explore ways to broaden their experience by gaining "requisite" knowledge in one or more fields in addition to their primary field.**

For example, graduate students can

*   Do this through master's theses or PhD dissertations that involve multiple advisers in different disciplines.
*   Share an office with students in other fields. Enhance their interdisciplinary expertise by participating in conferences outside their fields and in poster sessions that represent multiple disciplines. Those venues provide opportunities for junior researchers to present their work to colleagues outside their fields.

## Postdoctoral Scholars

**P-1: Postdoctoral scholars can actively exploit both formal and informal means of gaining interdisciplinary experiences during their postdoctoral appointments through such mechanisms as networking events and internships in industrial and nonacademic settings.**

For example, postdoctoral scholars can

- Seek formal and informal opportunities to communicate with potential research collaborators in other disciplines and develop a network of interdisciplinary colleagues.
- Broaden their perspective through internships in industrial settings or other nonacademic settings.

**P-2: Postdoctoral scholars interested in interdisciplinary work should seek to identify institutions and mentors favorable to IDR.**

For example, postdoctoral scholars can seek positions at institutions that

- Have strong interdisciplinary programs or institutes.
- Have a history of encouraging mentoring relationships across departmental lines.
- Offer technologies, facilities, or instrumentation that further one's ability to do IDR.
- Have researchers and faculty members with whom the postdoctoral scholar interacts who place a high priority on shared interdisciplinary activities.

## Researchers and Faculty Members

**R-1: Researchers and faculty members desiring to work on interdisciplinary research, education, and training projects should immerse themselves in the languages, cultures, and knowledge of their collaborators in IDR.**

For example, researchers and faculty members can

- Develop relationships with colleagues in other disciplines. Learn more about the knowledge and culture of other disciplines by participating in interdisciplinary projects.
- Actively seek opportunities to teach classes in other departments and give papers at conferences outside their own disciplines or departments. In their written and oral communications, researchers and faculty

members can facilitate IDR by using language that those in other disciplines are able to understand.

- Mentor students and postdoctoral scholars who wish to work on interdisciplinary problems.

**R-2: Researchers and faculty members who hire postdoctoral scholars from other fields should assume the responsibility for educating them in the new specialties and become acquainted with the postdoctoral scholars' knowledge and techniques.**

For example, researchers and faculty members can

- Familiarize themselves with the research cultures and evaluation methods of the postdoctoral scholars' fields.
- Learn about the career expectations of the postdoctoral scholars, when possible, and the demands that they will encounter in their careers.
- Guide the postdoctoral scholars toward interdisciplinary learning opportunities, including workshops, research presentations, and social gatherings.

### Educators

**A-1: Educators should facilitate IDR by providing educational and training opportunities for undergraduates, graduate students, and post-doctoral scholars, such as relating foundation courses, data gathering and analysis, and research activities to other fields of study and to society at large.**

For example, educators can

- Provide training opportunities that involve research, data-gathering, data analysis, and interactions among students in different fields.
- Demonstrate the power of interdisciplinarity by inviting IDR speakers, providing examples of major discoveries made through IDR, and highlighting exciting current research at the interfaces of fields.
- Encourage a multifaceted, broadly analytical approach to problem-solving.
- Include as part of foundation courses (such as general chemistry) materials that show how the subjects are related to other fields of study and to society at large.
- Show through explanatory examples the relevance of IDR to complex societal problems, which often require multiple disciplines and challenge current scientific and technical methods.
- Discourage the notion that some disciplines rank higher than others.

- Create more opportunities for students to learn how research disciplines complement one another by

— Developing policies and practices that support team teaching of interdisciplinary courses by faculty members in diverse departments or colleges.

— Modifying core course requirements so that students have more opportunities to add breadth to their study programs.

— Provide team-building and leadership-skills development as a formal part of the educational process.

# 5

# How Academic Institutions Can Facilitate Interdisciplinary Research

The previous chapter reviewed the environment and some of the challenges faced by individual researchers in approaching interdisciplinary research. This chapter reviews the opportunities and difficulties encountered by academic institutions that wish to facilitate IDR. Many institutions have become aware of institutional practices that create barriers to IDR; fewer have been able to lower or remove them. This chapter summarizes the barriers and describes how some institutions are trying to overcome them by reorganizing research, reallocating funds, and designing teaching programs conducive to interdisciplinarity.

## A VISION FOR INSTITUTIONS THAT WISH TO PROMOTE INTERDISCIPLINARY RESEARCH

Ideas for IDR may be generated from the bottom up, by individual researchers who want to cross disciplinary boundaries alone or in collaboration with others, or from the top down, by institutions and funding organizations that initiate and support research and teaching. This chapter discusses both approaches from the point of view of the institution.

In the committee's survey, respondents were asked to rank the general supportiveness for IDR at their current institution and up to two previous institutions on a scale of 0 (IDR-hostile) to 10 (IDR-supportive). There appears to be a trend toward more supportive environments for IDR, but it is also possible that respondents purposefully moved to institutions that were more supportive (Figure 5-1). There appear to be interesting relationships between general institutional supportiveness for IDR and both budget

## SURVEY

Institution's General Supportiveness for IDR from 0 (hostile) to 10 (supportive)

| Environment for IDR | Convocation Survey (n=91) | Individual Survey (n=423) | Provost Survey (n=57) |
|---|---|---|---|
| Current institution | 7.74 +/– 2.07 | 7.25 +/– 2.31 | 7.24 +/– 1.70 |
| Previous institution(s) | 5.95 +/– 2.17 | 6.35 +/– 2.57 | 5.67 +/– 2.04 |

FIGURE 5-1   Institutional environment for IDR.
NOTES: Respondents were asked to rank the general supportiveness for IDR at their current institution and up to two previous institutions on a scale of 0 (IDR-hostile) to 10 (IDR-supportive). Rankings are reported as mean +/– standard deviation. See Appendix E for more information on the three surveys.

and number of faculty. Respondents ranked their IDR experiences more favorably at institutions with budgets and faculty members at either end of the spectrum. This echoes findings by Epton et al.[1]

A vision of interdisciplinarity may begin with simple steps and behaviors that nourish the practice of collaboration. That might be done, for example, by creating more opportunities for faculty to work with students and postdoctoral fellows in different disciplines and departments. It might also be done by allocating seed money for space where a promising interdepartmental partnership can begin. One study notes that "interdisciplinary centers need not only to be well-funded but to have an independent physical location and intellectual direction apart from traditional university departments."[2]

Over half of the institutions represented in the committee's survey provided "venture capital" for interdisciplinary work. Amounts provided ranged from $1,000 to $1 million, but centered at $10,000-50,000 (Figure 5-2). Grant duration varied, but most tended to be 1- to 2-year awards.

A vision of interdisciplinarity might include a strategy to help young centers while they seek long-term support. For example, a university might give IDR high priority in its fund-raising and help to make the case with foundations to support an interdisciplinary strategy.

Or a vision might include a plan to broaden institutional participation wherein leaders can make the case for IDR through campus-wide meetings

---

[1]Epton, S. R., Payne, R. L., and Pearson, A. W. "Cross-Disciplinarity and Organizational Forms." *In:* Managing Interdisciplinary Research, Eds. Epton, S. R., Payne, R .L., and Pearson, A. W. Chichester: John Wiley & Sons, 1983.

[2]Rhoten, D. and Caruso, D. "Interdisciplinary Research: Trend or Transition?," *Items and Issues* 5(1-2):6, 2004.

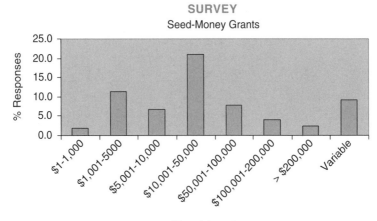

FIGURE 5-2   Size of seed-money grants.
NOTES: Respondents were asked whether their institution provided seed money to help start up interdisciplinary programs and were asked to briefly describe the amounts available and the major criteria used in making awards. Of the provost respondents, 87.7 percent indicated such awards were available. Interestingly, while 48.5 percent of the individual respondents answered yes, 27.2 percent were not aware of their institution's policy. Awards ranged from $1,000 to $1 million.

and discussions that air the needs of faculty and students.[3] Institutions can provide resources for curriculum development, student training in the use of equipment, and incentives for building synergies between IDR and teaching (see Box 4-2).

The top three recommendations for institutions from survey respondents were to foster a collaborative environment, to provide faculty incentives including hiring and tenure policies that reflect and reward involvement in IDR, and to provide seed money for IDR projects (Figure 5-3).

In practical terms, a vision might be implemented in many ways. To be effective, it would probably contain elements needed to overcome the barriers described in the next section.

---

[3]Roberts, J. A. and Barnhill, R. E. "Engineering Togetherness: An Incentive System for Interdisciplinary Research." ASEE/IEEE Frontiers in Education Conference. Reno, NV. October 10-13, 2001.

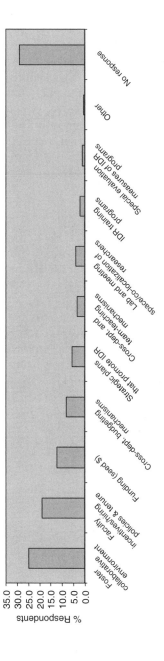

SURVEY

Institutions

**FIGURE 5-3** Recommendations for institutions.

NOTES: Survey Question: "If you could recommend one action that institutions could take that would best facilitate interdisciplinary research, what action would that be?" The top three recommendations for institutions (n = 341) were to foster a collaborative environment (26.5 percent), to provide faculty incentives (including hiring and tenure policies) that reflect and reward involvement in IDR (18.4 percent), and to provide seed money for IDR projects (11.1 percent). These recommendations and those for departments (Figure 5-5) reflect preconditions for IDR and are mirrored in recent self-studies carried out at the University of Washington (2002), University of Michigan (2000), and University of Michigan (1999), and by the American Association of Colleges of Pharmacy (2002).

SOURCE: Klein, J. T. and Porter, A. L. "Preconditions for Interdisciplinary Research." *In:* International Research Management Studies in Interdisciplinary Methods from Business, Government, and Academia, Eds. Birnbaum-More, P. H., Rossini, F. A., Baldwin, D. R. New York: Oxford University Press. 1990. pp. 11-19.

## INSTITUTIONAL BARRIERS TO INTERDISCIPLINARY RESEARCH

Even the most supportive leadership must contend with substantial barriers that impede IDR. The committee's surveys suggested widespread awareness of barriers to IDR: 71 percent of respondents to the Individual Survey and 90 percent of respondents to the Provost Survey reported a belief that major impediments to IDR existed in their institutions. Barriers often stem from customs that have evolved over many decades, generally for sound reasons. It is ironic that some of the barriers are consequences of an otherwise excellent academic system that supports frontier research at every level and achieves great depth in training future generations of scientists. As shown by the boxed examples throughout this report, however, many institutions have developed practical ways to reduce the impediments.

### Limited Resources

Of course, time and resources devoted to facilitating IDR are diverted from existing activities (both interdisciplinary and disciplinary). Starting a new program, providing new seed funds, or creating a new IDR center often means closing or reducing an effort in another area.

As a result, the institutional leadership needs to evaluate proposals for new activities carefully to ensure that they are not just satisfactory but outstanding. Some key mechanisms for doing so are to focus resources on activities with long-term implications and to involve high-quality senior faculty and promising junior faculty. Institutional leaders may also wish to establish an advising committee of faculty successful in IDR to evaluate proposed new activities; they are knowledgeable and likely to be sympathetic, and yet they are competitors for the same funds.

### The Academic Reward System

Traditional academic systems for hiring, tenure, promotion, space allocation, and other rewards may constitute a substantial barrier to IDR (see Figure 4-5). At most academic institutions, hiring, tenure, and promotion are controlled by departments, and faculty often receive credit only for the teaching and research actually performed in their departments. Faculty who teach in interdisciplinary teams or classes outside the department may receive little or no departmental credit.

### Different Institutional Cultures

Differences in culture—a set of customs, shared values, understandings, and relationships that pervade a discipline or unit—slow the communication and cooperation that underlie IDR.[4] The culture of a mathematics department, for example, differs in many ways from that of a biology department; potential collaborators may have to work hard to agree on such concepts as "proof" and "precision."

---

**Convocation Quote**

Most institutions have scientists in discrete departments, and while there are some enlightened institutions, there are many where if you are in biology, you are not allowed to speak to those nasty folks in chemistry, much less to sociologists, who are someplace else and you wouldn't know what to say to them even if you met them.

Lawrence Tabak, director of the National Institute of Dental and Craniofacial Research, National Institutes of Health

---

### Program Evaluation

Traditional program evaluation evolved to review departments and associated education and training programs. A quick look at the listing of science, engineering, and humanities fields used by the National Science Foundation in its Survey of Earned Doctorates, for example, shows little change over the 40 years that the survey has been performed, 1960-2002. The same is true of the National Research Council in its Assessment of Research Doctorates. Academic institutions rely on such data to benchmark their programs and allocate internal resources (see Box 5-1). When emerging fields are not included in assessments, academic institutions tend to leave them out of the resource allocation as well.

Survey respondents were asked to describe evaluation methods used by their institutions to evaluate interdisciplinary programs. The predominant ones cited were internal and external visiting committees and informal feedback (see Figure 5-4).

### Different Departmental Policies and Procedures

Departments and other units often balk at collaboration because of different administrative customs. Departments commonly differ over

- Allocation of indirect-cost recovery funds.

---

[4]Feller, I. "Whither interdisciplinarity (In an Era of Srategic Planning)?" Presented at AAAS Meeting, Seattle, WA, Feb. 15, 2004.

EVOLUTION

**BOX 5-1   Assessing Research-Doctorate Programs[a]**

Some researchers have questioned the reinforcing role of the National Research Council "rankings" on the "stiffness" of disciplinary boundaries and wondered whether and how new fields can or will be included in upcoming assessments. Certainly, university administrators pay great heed to the NRC assessments, and many base resource allocations—not to mention recruitment strategies—on them. Given the importance of the assessments, there is concern that emerging fields and extradepartmental programs, many of which are interdisciplinary, be included.

Partially in response to those concerns, the NRC recently completed a study to decide whether and how another assessment of research-doctorate programs should be conducted. The committee charge was as follows: "The methodology used to assess the quality and effectiveness of research doctoral programs will be examined and new approaches and new sources of information identified. The findings from this methodology study will be published in a report, which will include a recommendation concerning whether to conduct such an assessment using a revised methodology." The committee was informed through the deliberations of panels that considered taxonomy and interdisciplinarity, quantitative measures, student processes and outcomes, and measures of reputation and data presentation.

The committee concluded that undertaking the assessment again would be valuable and made specific recommendations with regard to taxonomy and inter-

- Organizing research and teaching.
- Allocating credit for multiauthor papers, especially when authors are in different disciplines or institutions.
- Control of space or capital-intensive facilities.
- Agreement on standards for recruiting and evaluating faculty with joint appointments.[5]

Among the top recommendations for departments listed by survey respondents were adopting new organizational approaches, recognition of faculty and researchers for interdisciplinary work, and adapting departmental resources and support for IDR (see Figure 5-5).

### Lengthy Startup Times

Some kinds of research programs, and especially IDR, require long

---

[5]Feller, I., ibid., pp. 10-11.

disciplinarity concerning which fields and which programs within fields should be included in the study. While most of the criteria for inclusion of fields used in the earlier (1995) report were retained, it had new recommendations on the identification and listing of emerging and interdisciplinary fields. In particular, emerging fields should be identified based on the basis of their increased scholarly and training activity. The number of programs and degrees may not be sufficient to warrant full-scale evaluation at this time. If possible, emerging fields should be listed as subfields; otherwise they should be listed separately.

To gather data on programs and faculty, the committee recommended graduate programs be asked to identify those interdisciplinary centers within which their graduate students conduct research. Faculty would be asked to identify all the programs in which they taught graduate courses or supervised dissertations.

Finally, the report recommended some changes in broad fields and the inclusion of sub-fields to assist programs in placing themselves in the taxonomy:

• Fields should be organized into four major groups rather than the five of the previous Research Council study. Mathematics and physical sciences should be merged into one major group with engineering.
• Biological sciences, one of the four major groups, should be renamed "life sciences."
• Subfields should be listed for many of the fields.

[a]More information on the National Academies report "Assessing Research-Doctorate Programs: A Methodology Study (2003)" can be found at: *http://books.nap.edu/catalog/10859.html*.

startup times to arrange equipment, staffing, or infrastructure. Participants must delve deeply into another language and culture at the outset of a project.[6] Yet the policies and procedures specified by funding organizations and major universities do not always accommodate that need. The extra time required for IDR, even if well spent, can lead to fewer substantive results and publications, but the tenure and funding clock is not calibrated to take such activities into account.

### Decentralized Budget Strategies

Most of the traditional academic budget is allocated to recurring categories, such as salaries, physical-plant costs, and instructional expenses. Flexible funds tend to be assigned to departments and colleges as operating

---

[6]Bruhn, J. G. "Interdisciplinary research: A philosophy, art form, artifact or antidote?" *Integrative Physiological and Behavioral Science*, Vol. 35, No. 1, January-March 2000, p. 62.

**SURVEY**

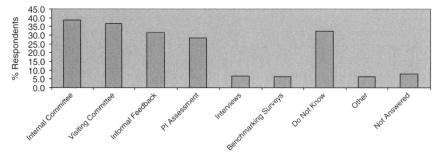

FIGURE 5-4  Institutional methods for program evaluation.
NOTES: Respondents were asked to describe dominant forms of evaluation used by their institutions to evaluate interdisciplinary programs. Institutions used multiple forms, the predominant methods being internal and external visiting committees, informal feedback, and PI assessment. Trends in evaluation methods reported by individuals and provosts were similar, but 37 percent of individual respondents were not aware of institutional evaluation policies.

**SURVEY**

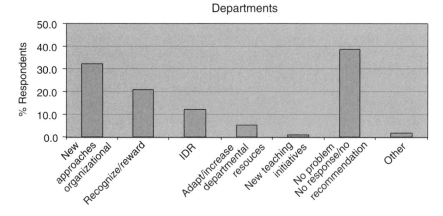

FIGURE 5-5  Recommendations for departments.
NOTES: Survey Question: "If you could recommend one action that departments could take that would best facilitate interdisciplinary research, what action would that be?" The top three recommendations for departments (n = 294) were to adopt new organizational approaches to IDR (32.1 percent), to recognize and reward faculty and other researchers for interdisciplinary work (20.8 percent), and to adapt or increase departmental resources to support IDR (12.3 percent).

units. As a result, central administrations often have scarce fiscal resources for initiating or sustaining IDR programs. Departments may be reluctant to contribute resources for activities not seen as directly beneficial.[7]

## A NEED FOR SYSTEMATIC INSTITUTIONAL REFORM

The overall effect of barriers is hard to quantify, but even slight deterrents to researchers who are trying to reach career milestones—such as earning a degree, locating an academic position, raising funds, attaining tenure, publishing the results of research, or sustaining a long-term research portfolio—can become substantial and even onerous in the aggregate. This "accumulation of disadvantage," or theory of limited differences, has been discussed extensively in recent years, particularly as related to the disadvantages of women and members of other underrepresented populations in science and engineering.[8]

Many, perhaps most, universities are aware of the adverse effects of the barriers to IDR. Some have described reforms, placed them in strategic plans, and even allocated money for new initiatives. Few universities, however, have implemented systematic reforms to lower institutional barriers. A study that examined the interdisciplinary centers of major universities reported that "universities are failing to 'walk the walk'—or even to comprehend fully what doing so would entail."[9]

Still, some universities have begun to implement reforms, and it is on these new experiments and procedures that the present report focuses. As suggested in Chapter 4, the needs of students, postdoctoral fellows, and faculty change as they advance through the stages of a research career. The suggestions and examples in the next section are organized to reflect the progression of needs.

---

[7]González, C. "The Role of the Graduate School in Interdisciplinary Programs: The University of California, Davis, Budget Model," Council of Graduate Schools *Communicator*, June 5, 2003.

[8]The concept of "accumulation of advantage and disadvantage" is discussed by Cole, J. R. and Singer, B. "A theory of limited differences: Explaining the productivity puzzle in science," in Zuckerman, H., Cole, J. R., and Bruer, J. T. eds., *The Outer Circle: Women in the Scientific Community*, New York: W.W. Norton, 1991, pp. 277-310; Merton, R. K., "The Matthew Effect in Science," *Science* 159, No. 3810, January 5, 1968, pp. 56-63; Zuckerman, H. A. *Scientific Elite: Studies of Nobel Laureates in the United States*, New York: The Free Press, 1977; and Sonnert, G. "Who Succeeds in Science? The Gender Dimension." Rutgers University Press, 1995.

[9]Rhoten, D. and Caruso, D. *Lead, follow, get out of the way: sidestepping the barriers to effective practice on interdisciplinarity*. The Hybrid Vigor Institute, April 2001, p. 4. Available at: *http://www.hybridvigor.net/interdis/pubs/hv_pub_interdis-2001.04.30.pdf.*

INNOVATIVE PRACTICE

**BOX 5-2   Breaking Down Institutional Barriers by
Breaking Bread Together**

One of the overarching themes in facilitating interdisciplinary research is finding ways to bring together researchers who would not otherwise meet. In the committee's survey, the top recommendation was for institutions to foster a collaborative environment—to provide opportunities for interaction across disciplines and
allow greater movement of faculty among programs and departments. That theme
was echoed in recommendations to funding agencies, professional societies, and
researchers themselves. One director called IDR "a body-contact sport—people
have to be running into each other to make it work."[a]

To that end, several academic institutions have designed research centers
with architectural features that promote collaboration, from cafeterias to shared
laboratory space. As one director emphasized, "The last thing that I am going to
shut down in my building is the cafeteria. It is tremendously important to bring
people out of their buildings, out of their offices, out of their labs, and into a common space, and then they start talking."[b]

Even in industry, where laboratories are usually organized in interdisciplinary
teams, common areas are important. "There is something about breaking bread
together that causes creative juices to flow. If you go into our cafeteria at lunch-
time, you find lots of interactions occurring. We have set up conference rooms
around our cafeteria so that people can walk in there and start writing on easels or
white boards or whatever. We promote collaborative work. We promote it because
it is a way of life for us. It is what provides our bread and butter."[c]

At the other end of the cost spectrum is providing space for regular meetings
of researchers across disciplines, departments, and colleges. "Despite the age of
high technologies and computer communications, rubbing shoulders really still

## FACILITATING INTERDISCIPLINARY RESEARCH
## AND EDUCATION

There is considerable overlap in activities between researchers at different stages of a research and teaching career, and the structure of this section
is not intended to create artificial divisions. In fact, the concerns and goals
of a student may be quite similar to those of a faculty member. For example, as indicated above, both students and faculty who wish to do IDR
face difficulties in learning the language, culture, and knowledge of other
disciplines.[10] Institutions can take the lead in providing incentives for stu-

---

[10]Metzger, N. and Zare, R. N. "Interdisciplinary research: From belief to reality," *Science*
283(5402):642-643, 1999.

helps."[d] Something as simple as providing institutional support for use of a meeting room can be pivotal in assembling a team. The Fred Hutchinson Cancer Research Center goes one step further and supports an "interdisciplinary club,"[e] which brings together graduate students, postdoctoral scholars, and staff researchers to discuss research ideas.

Funding organizations can help by providing venues or funding for meetings to discuss interdisciplinary topics. For example, the National Academies Keck Futures Initiative (NAKFI), sponsors annual conferences, to which about 100 scientists from different research settings are invited to participate in discussions centered on an emerging cross-disciplinary research theme.[f] As mentioned earlier in this report (see Box 2-4), dispersion, rather than multidisciplinarity, is often the most problematic aspect of interdisciplinary projects. Mechanisms that bring researchers together are effective in increasing project success.

---

[a]Jeffrey Wadsworth, director, Oak Ridge National Lab. Comments made at Convocation on Facilitating Interdisciplinary Research, Washington, D.C., January 29, 2004.

[b]Pierre Wiltzius, director, Beckman Institute for Advanced Science and Technology, University of Illinois at Urbana-Champaign. Comments made at Convocation on Facilitating Interdisciplinary Research, Washington, D.C., January 29, 2004.

[c]Uma Chowdhry, vice president, Central Research and Development DuPont. Comments made at Convocation on Facilitating Interdisciplinary Research, Washington, D.C., January 29, 2004.

[d]Harvey Cohen, Professor of Pediatrics, Stanford School of Medicine and chair, Interdisciplinary Initiatives Program. Comments made at Convocation on Facilitating Interdisciplinary Research, Washington, D.C., January 29, 2004.

[e]Paulson, T. (2003) Grassroots Interdisciplinary Training: The FHCRC Interdisciplinary Club. Science's Next Wave, January 3, 2003 http://nextwave.sciencemag.org/cgi/content/full/2002/12/30/7.

[f]Keck Futures Initiatives Web site http://www7.nationalacademies.org/keck/Keck_Futures_Conferences.html.

---

dents and researchers to interact with other disciplines and to learn other languages and cultures (see Boxes 4-1 and 5-2).

Similarly, both students and faculty benefit from lowering the barriers to team teaching of interdisciplinary courses. For students, the exposure to teachers in different disciplines can lead to understanding that is broader than a single discipline. For faculty, the ability to collaborate with teachers in different disciplines may lead to new understandings of their own and an ability to describe their work to students in different majors. Institutions facilitate both research and teaching when they support team teaching through better methods to recognize and reward teachers who are teaching outside their departments, through teaching-credit policies that sustain team-taught courses, through opportunities for students to acquire mentors in multiple disciplines and with different perspectives, and through stronger support for departments engaged in team teaching (see Box 4-2).

**Convocation Quote**

We have had to put a lot of care into how we are a community and what we do to keep that growing. I think that as we aged a little bit and we had more people involved—more students and more faculty mentors—we suddenly hit "critical mass." There was a big difference; there was momentum.

Marye Ann Carroll, professor,
Atmospheric, Oceanic, and Space Sciences, Univ. Michigan

The intent of this section is not to differentiate, but to point to common themes. To oversimplify somewhat, institutions can best facilitate IDR by considering the drivers of IDR discussed in Chapter 2: creating collaborations capable of addressing the enormous complexity of nature, allowing students and faculty the flexibility to explore the interfaces between disciplines, extending partnerships to the humanities and other sectors required to address complicated societal problems, and providing access to and understanding of the "generative technologies" whose full exploitation may lead to new fields and new ways of looking at existing fields.

### Undergraduate Education

Undergraduate students often show great enthusiasm for interdisciplinary and problem-driven questions, including those of societal relevance. There are many ways in which institutions can design undergraduate (or even high school) programs that take advantage of that natural interest:

- Undergraduate interdisciplinary degree programs: The number of interdisciplinary undergraduate majors has begun to grow in recent years, and numerous models are now available.
- Undergraduate research programs: The variety of research experiences for undergraduates (REUs) is increasing rapidly, and students have responded with strong interest.[11]
- Topics of high societal relevance: Offering courses or programs on such topics may attract a different mix of students, including those who want to perform research of practical use.
- Programs that offer depth in more than one discipline: Multiple skills can be developed by a broader training program, including studies and internships in other fields, exercises in combining approaches of mul-

---

[11]At the University of Michigan, students participating in REUs have higher rates of retention in science and engineering. 1996 Assessment of the Undergraduate Research Opportunity Program. Available at *http://www.undergraduate.research.umich.edu/homeassessUROP.html*.

---

### INNOVATIVE PRACTICE

## BOX 5-3   IDR at Primarily Undergraduate Institutions[a]

Undergraduate research is a growing phenomenon. The Council on Undergraduate Research (CUR),[b] founded in 1978, supports and promotes high-quality undergraduate student-faculty collaborative research and scholarship. CUR has 3,000 members representing over 870 institutions in eight academic divisions. Much of this research is interdisciplinary. For example, Haverford College, an undergraduate institution with about 1,100 students, is on the cusp of a major change in curriculum. Its plan for the next 5-10 years is to do away with general courses in chemistry, physics, and biology and to teach them integratively. The idea is to teach chemistry and physics as an integrative course in the first year, providing foundations for further work in the disciplines and a foundation for an integrated course in organic chemistry and molecular biology. The first 2 years of the curriculum would emphasize mathematics and statistics.

In the junior and senior years, there is already a fairly broad curriculum that is taught in an interdisciplinary way. Juniors in the chemistry, biology, and physics departments take introduction-to-research-methods courses instead of traditional laboratory courses. These intensive courses last for the entire school year.

In the senior year, students are immersed in research. That is, research is integrated into the curriculum: students are introduced to research methods instead of having to learn physical and chemical laboratory methods, inorganic and organic chemistry, and so on. All these concepts are pulled together into a single laboratory course, which is going to be expanded by units on material science, computational biology, neuroscience, and biophysics, in which students will navigate from module to module across the involved departments. The plan is to weave research and interdisciplinary work completely into the fabric of the curriculum of all the science departments.

---

[a]Julio de Paula, Professor of Chemistry, Haverford College, comments at the Convocation for Facilitating Interdisciplinary Research, Washington, DC, January 30, 2004. *http://www7.nationalacademies.org/interdisciplinary/Convocation_Agenda.html.*
[b]*www.cur.org.*

---

tiple disciplines, communication, and opportunities for portable scholarships, summer laboratory jobs, and industrial internships.

### Graduate Education

While graduate students are building a firm base in their primary discipline, they may become familiar with additional fields or skills that can extend their knowledge. To facilitate the ability of graduate students to ground themselves in interdisciplinary thinking, institutions can provide

- Programs with many of the same general features as undergraduate interdisciplinary courses but with added complexity and depth.
- Additional exciting research at the interfaces of disciplines, including opportunities to work with and learn from graduate students in other disciplines and multiple advisers who bring diverse perspectives to research problems.
- Additional academic recognition and funding that allow graduate students in IDR to anticipate prospects for advancement equal to those of single-discipline students.
- Graduate IDR internships, including assistance in finding appropriate academic "homes"; these are needed when departments are unable or unwilling to accommodate researchers doing interdisciplinary work.
- Experience in using instrumentation and other techniques that are beyond the inventory of a single adviser or discipline.
- Dual mentors who may bring different perspectives to the same problem.

### Postdoctoral Fellowships

For postdoctoral scholars, there is no substitute for honing expertise in one discipline; even researchers who direct interdisciplinary teams prefer members who are expert in at least one field rather than "masters of none." At the same time, many postdoctoral scholars are ready to benefit from complementary expertise in another field. Institutions can enrich the postdoctoral experience by providing

- Opportunities to interact with specialists in other disciplines and to learn the language, culture, and knowledge of a new discipline.
- Scholarships for gaining a master's degree in another field.
- Attentive mentoring by multiple mentors, with annual reviews so that postdoctoral scholars do not "fall through the cracks."
- Access to a broader array of instrumentation and analytical techniques.
- Appropriate referees and mentors who will support the inclusion of IDR in tenure decisions.
- Opportunities to undertake study in a foreign country.

### Hiring

Interdisciplinary faculty hiring requires changes that start long before the candidate is hired during the search and interview processes. Most search committees reside within a department or discipline. If interdisciplinary search committees are formed, successful searches require that the relationship between these and the departmental search committees be agreed on and well understood. Interview schedules, also often the responsibility of department committees, must cut across departments. Department administrators must negotiate terms of joint appointments, including startup resources and space. Institutions have experimented with ways to lower the barriers to hiring junior scientists working in IDR that were described in Chapter 4. Some have adopted institutionwide hiring policies that promote IDR (see Box 9-5). Others have provided transitional funding for hiring interdisciplinary people. The University of Wisconsin uses a "cluster hiring" program (see Box 5-4), and Arizona State University has split departmental appointments for more than a decade.

Here are other examples:

• Columbia University has allocated 15 faculty lines, mostly to junior faculty, agreeing to pay salaries for the first 5 years with departments to assume support thereafter. This incentive program is funded by intellectual-property revenues.
• The National Center for Atmospheric Research reserves four slots per year to hire assistant professors with interdisciplinary interests. The institution and the departments each provide half the support.
• The California Institute of Technology has plans to hire about 25 interdisciplinary faculty in information technology.

Research institutions also have increasing needs to hire and provide a career track for scientific managers, as recommended in the National Research Council report on team science.[12] The managers, in turn, would need thorough interdisciplinary training.

### Junior Faculty

Junior faculty can benefit from many of the same research opportunities as postdoctoral scholars. In addition, modest institutional changes can help them to overcome departmental or professional barriers:

• Institutional funding for junior faculty positions can include more flexible teaching placement.
• The work done by faculty in interdisciplinary centers or team-teaching situations should count with equal credit toward promotion and tenure.

---

[12]National Research Council. 2003. *Large-Scale Biomedical Science: Exploring Strategies for Future Research.* Eds. Nass, S. J. and Stillman, B. W. Washington, D.C.: The National Academies Press.

---

**INNOVATIVE PRACTICE**

**BOX 5-4   The Cluster Hiring Initiative
at the University of Wisconsin**

The Cluster Hiring Initiative (CHI)[a] at the University of Wisconsin at Madison (UW) grew out of the campus strategic planning process of the middle 1990s. The initiative involved a provost-coordinated campuswide competition to identify groups of new faculty hires, or "clusters," to work together on interdisciplinary programs and emerging fields of inquiry.

By establishing the CHI, the campus acknowledged that existing curriculum demands, department traditions, and faculty governance may limit department opportunities to pursue new directions in faculty hiring. Departments may be unable to hire faculty who pursue important new, more experimental, less established lines of research or interdisciplinary research that is by definition distant from the core of a single discipline. The prevailing academic cultures and structures tend to replicate existing areas of expertise, reward individual effort rather than collaborative work, limit hiring input to a single department in a single school or college, and limit incentives and rewards for interdisciplinary and collaborative work.

The provost invited proposals from faculty that identify promising subjects for faculty collaboration. Since 1998, faculty have submitted hundreds of proposals to fund faculty lines to pursue and develop new and promising areas of interdisciplinary and collaborative inquiry. These are permanent lines that remain with the hiring department as long as a cluster faculty remains with the university. The campus has conducted five phases of cluster identification and funding. Through 2003, 49 clusters with 137 new faculty lines were authorized with central funding, and schools and colleges matched six additional cluster faculty positions.

The provost-appointed Faculty Advisory Review Committee is composed of one person from each of the four divisional and research committees and two at-large members appointed by the chancellor. Coordinated by the assistant vice chancellor for faculty and staff programs, the committee evaluated preproposals and full proposals against five criteria[b]: quality and merits of the initiative, relevance to the mission and vision of UW, timing, potential for success, and potential for faculty diversity.

Some departments have used cluster positions to add to or strengthen their department core disciplines. In other cases, clusters strengthened existing interdisciplinary programs. An evaluation committee[c] heard more enthusiasm than criticism about the promises and activities of the initiative. However, faculty expressed concerns about tenure review, salary equity, and infrastructure support:

---

• Faculty should receive equal credit at their home institutions for contributions to interdisciplinary or multidisciplinary journals or conferences.

• Faculty can be permitted to request reviews in other fields at the third year and to request review panels that include extradepartmental expertise (see Box 5-5).

*Tenure Review:* Departments and divisional committees found it difficult to evaluate a cluster faculty's interdisciplinary scholarship during the tenure review. However, an ad hoc interdisciplinary committee report[d] showed "no difference in the likelihood of achieving tenure among probationary faculty with multiple appointments compared to those with appointments in only one department." In fact, data from this ad hoc report showed that "the likelihood of achieving tenure is not lower but in fact higher for candidates with joint appointments or multiple tenure homes." A similar concern about documenting scholarship for tenure was expressed recently by the campus clinical faculty; however, a committee that examined this concern found no evidence that clinical faculty were achieving tenure at a lower rate than other faculty in the health sciences.[e]

*Salary Equity:* The committee did find that CHI appointments across school and college lines have increased faculty awareness that courses taught, salaries, and startup packages differ widely with the field and area of specialization. As cross-college and cross-department appointments increase, the campus may need to pay more attention to merit processes that involve input from schools, colleges, and departments with which cluster faculty are involved.

*Infrastructure Support:* In response to the identified need to foster cluster infrastructure, the provost established a campuswide Cluster Hiring Enhancement Grant competition to provide partial support for graduate students, program assistants, and laboratory assistants and other expenses related to programmatic activities.

---

[a]Cluster Hiring Initiative Program Description, Office of the Provost, University of Wisconsin. Homepage *http://wiscinfo.doit.wisc.edu/cluster/progrmdesc.html*. Accessed April 30, 2004.

[b]Cluster Hiring Initiative Program Overview and Guidelines, Office of the Provost, University of Wisconsin. *http://wiscinfo.doit.wisc.edu/cluster/overviewr5html*. Accessed April 30, 2004.

[c]*Report of the Provost's Ad Hoc Advisory Committee to Evaluate the Cluster Hiring Initiative University of Wisconsin-Madison.* Submitted to the Provost, November 11, 2003. (Coordinator, Linda Greene, Associate Vice Chancellor for Faculty and Staff Programs).

[d]The provost's Ad Hoc Committee on Faculty in Interdisciplinary Programs, chaired by Elizabeth Thomson, appointed by the provost to identify potential disparities in responsibilities and rewards between faculty with interdisciplinary responsibilities and those without, submitted its report to the provost on March 8, 2003.

[e]*Report of the Health Sciences Division Task Force on the Health Sciences Division Proposal* submitted to the Deans of the Health Sciences Schools, April 23, 2003 (Chair Professor John Mullahy, Dept. of Population Health Sciences) Appendix F, pages 75-76 and Appendix I pages 89-90.

## Tenured Faculty

Tenured faculty are often more active in IDR than junior faculty because their career positions are secure. But institutions can help senior faculty through several modest policy changes:

- Developing incentives that allow faculty to continue their education in fields complementary to their own.
- Creating mechanisms for interdisciplinary work or projects to be evaluated by panels on which multiple disciplines are represented.
- Providing more opportunities for faculty to learn from students and postdoctoral scholars in other fields.
- Using seed money to fund sabbaticals and visiting-scholar grants for faculty to work in multidisciplinary groups.

---

**Convocation Quote**

Rockefeller University really understands what research is about. Research is focused on a problem. You find the tools, solve the problem. So, a year after I was hired, they asked me, "By the way, what is your title?" That is the appropriate response to a professor. Let the professor tell you what he or she wants to do.

Joel Cohen, Abby Rockefeller Mauze Professor, Laboratory of Populations affiliated with both Rockefeller University and Columbia University

---

### All Faculty

Some of the most important reforms that institutions can undertake apply to both junior and senior faculty. They include these:

- Reward structure: Faculty who conduct IDR need professional recognition comparable with that given to faculty who conduct single-discipline research.
- Faculty evaluations: Academic leaders can make special efforts to overcome departmental or disciplinary bias in reviewing (see Box 5-5). Faculty are treated fairly when they are evaluated on the basis of all their work—not just the work in the discipline of their home departments.
- Publication credit: Faculty benefit by receiving institutional credit for work reported in journals or conferences outside their specialties or in interdisciplinary journals.
- Allowance for long startup times: Universities can be flexible with respect to time in their tenure-review processes or allow longer probationary time for nontenured faculty when some or all of their contribution is interdisciplinary.
- Curricular integration: A curriculum that allows formal placement of IDR on the teaching agenda provides a strong, visible endorsement.

---

TOOLKIT

## BOX 5-5    Providing for Interdisciplinarity in the Tenure and Review Process[a]

At the University of Southern California (USC), IDR gained prominence in 1994, owing to a new strategic plan that called for the development of undergraduate research programs focused on IDR.[b] "Ideas can bubble up from the bottom, but they need to be embraced by the top," explained Neil Sullivan, USC vice provost for research. Sullivan's primary responsibility is to facilitate multidisciplinary research across the university.

Several mechanisms have been put into place to encourage IDR:

• *Research and Incentive Fund:* For inhouse peer-reviewed proposals for projects from more than two faculty members in more than two schools of the university.
• *Faculty Fellowships:* Up to $50,000 for IDR proposals and release from teaching. Proposals are reviewed by other faculty members at the university. Awardees meet monthly to make presentations and give progress reports, and their advice is solicited by the vice provost on ways to break down barriers to IDR.
• *Specific Guidelines:* USC has added explicit language in its promotions and review criteria for interdisciplinary scholarship and teaching, and IDR was specifically addressed in the provost's cover letter with the guidelines.[c] Within the guidelines, specific points address IDR:

> If a candidate's scholarship is interdisciplinary, the department and school should take special care to evaluate the work properly. If work does not match the departments' priorities, but does further the school or University policies, that should be explained. The evaluation of quality and quantity should be distinguished from discussion of how the work fits strategies for excellence.

Regarding selection of referees, the guidelines state:

> For interdisciplinary scholarship, the lists of external referees should include experts from the other discipline, as well as experts in the individual's own type of interdisciplinary scholarship.

---

[a]From an interview with Cornelius Sullivan, Vice Provost for Research, USC, November 10, 2003.
[b]University of Southern California 1994 Strategic Plan: *http://www.usc.edu/about/stra tegic_priorities/strategic_plan94.html.*
[c]"Guidelines of the University Committee on Appointments, Promotions, and Tenure." University of Southern California, issued October 27, 2003. Available on-line at *http://www. usc.edu/policies.*

## Institutional Leadership

Promoting IDR often begins with the central administration. Presidents, provosts, vice presidents for research, and other leaders have high visibility and good access to resources. According to the literature, the more open a person is to new experiences, the more creative he or she is likely to

TOOLKIT

**BOX 5-6    The Beckman Institute at the University of Illinois, Urbana-Champaign[a]**

The origin of the Beckman Institute at the University of Illinois is a story of interdisciplinarity. In 1983, the vice-chancellor for research appointed two faculty committees—one in the physical sciences and engineering, the other in the life and behavioral sciences—to explore the prospects for a radically new, campus-based research institute that drew on university disciplinary expertise and interests.

UIUC presented Dr. and Mrs. Arnold Beckman with the committees' proposal to create in a state-of-the-art institute an integrated array of research efforts that would be a model of interdisciplinary research. In 1985, the Beckmans awarded the University of Illinois $40 million for construction of the institute, and the state of Illinois made added commitments. The institute, a 300,000-ft$^2$ facility, began operations in early 1989. Many special features novel for academic settings and intended to foster interactions were incorporated into its design.

Faculty Affiliations and Reporting Structures: The director of the institute has the status of a dean and reports to the provost. All faculty in the institute have appointments in departments and maintain departmental teaching and service obligations. Some faculty are full-time; that is, all their research activities are centered in the institute. Others are part-time; they maintain some research space in the institute and some in departments. Still others have looser affiliate appointments; they are involved in an institute program and may have students or post-doctoral fellows working there, but they do not maintain offices. About 130 faculty are affiliated with the institute, with some 400-450 graduate students, 200-300 undergraduate students, and 70-80 postdoctoral fellows. A staff of 60-70 provide technical and administrative support.

Research Programs and Evaluation: The institute is organized along themes that cross-cut and build on university strengths in the physical sciences, engineering, and the cognitive and social sciences. Each of the major research themes is

be. That implies that openness should be taken into consideration when selecting a person to head an interdisciplinary education or research program if it is to be effective.[13] It is up to institutions to recognize innovative,

---

[13]Feist, G. J. and Gorman, M. E. 1998. The Psychology of Science: Review and Integration of a Nascent Discipline. *Review of General Psychology* 2, no. 1:3-47; Simonton, D. K. 2004. *Creativity in Science: Chance, Logic, Genius, and Zeitgeist.* New York: Cambridge University Press; Simonton, D. K. 2003. Scientific Creativity as Constrained Stochastic Behavior: The Integration of Product, Person, and Process Perspectives. *Psychological Bulletin* 129, no. 4:475-94.

evaluated every few years with the help of external experts. Is the work being done of the highest caliber? Is the research of individual faculty or groups of faculty taking advantage of the uniqueness of the institute? Is it interdisciplinary? When a review is unfavorable, the director has the duty to require faculty or groups of faculty to leave the institute and return to their home departments. The review process is important in the success of the institute. Turnover of research programs and individuals is essential to the institute's long-term vitality.

Relations with Departments and Colleges: Because the institute stands apart from the traditional college and departmental organization of the university, its relations with departments and colleges require continuing attention. Campus policy provides for sharing of indirect cost returns (ICRs) on grants with colleges and departments. For grants that involve a single investigator, or a group of investigators from a single department, the ICRs will all accrue to the home department, even though the work was performed in the institute. However, the ICRs on multi-investigator grants involving faculty from different departments pass to the institute. That rule has occasioned some controversy, especially regarding large grants involving many faculty.

The institute participates actively in the recruitment of new faculty when a department's interests intersect productively with those of Institute programs. Funds for equipment, student support, and other research needs are regularly allocated from those available in the institute and departments and colleges negotiate over how faculty allocate their time and interests to departmental and collegial affairs as opposed to institute affairs.

---

[a]Pierre Wiltzius, Director, Beckman Institute for Advanced Science and Technology, and Professor, Materials Science and Engineering Department and Physics Department, University of Illinois at Urbana-Champaign. Based on comments at the Convocation on Facilitating Interdisciplinary Research, Washington, DC. January 29, 2004. Beckman Institute for Advanced Science and Technology at the University of Illinois at Urbana-Champaign, Web site http://www.beckman.uiuc.edu/.

flexible leaders and to encourage them to take risks in discerning and supporting fresh ideas.

## Incentives and Rewards

One cause of turf battles between departments is that deans, department chairs, and other administrators are rewarded for strengthening their own departments, not for building links to others. Institutions can reward leaders for initiating interdisciplinary programs and can provide incentives for departments to share indirect cost revenues, seed money, course-credit assignments, intellectual property, space, personnel, and other resources[14] (see Box 5-6).

## Promoting Interactions

Institutions can also facilitate the natural development of departments as their researchers continually seek interaction with other disciplines.[15] Good leadership can assist interdepartmental interactions, which are often hindered by organizational structures (see Box 5-7). In particular, increased interaction with those outside one's department should be rewarded through the promotion and tenure process.

Biology, for example, has developed extensive interactions with mathematical science; this reflects the discipline's need for powerful quantitative tools. Despite that development, the two disciplines remain largely distinct at the institutional level. Often, the same barriers that hold back IDR hold back the natural evolution of the disciplines themselves.

---

**Convocation Quote**

Keeping a team motivated through ups and downs and through years of striving because nobody has done this type of work before takes a lot of . . . emotional intelligence. It takes understanding human behavior. It takes understanding human interactions and what keeps people motivated.

Uma Chowdhry, vice president for
Central Research and Development, DuPont

---

## Budget Reforms

Most major universities have developed decentralized budgeting models in which the lion's share of resources flows to schools, departments, and other units. This leaves relatively few resources to be used for "the com-

---

[14]An example of such a policy could be seen until recently at the Massachusetts Institute of Technology, where the administration allowed the use of the old Building 20 as a home for new, often interdisciplinary.

[15]As Blau has written, "The distinctive departmental structure of American universities makes it relatively easy to offer positions to specialists in new fields who work at the frontiers of knowledge, at first within departments and later, as the specialty grows, by establishing a separate department for it." Blau, P. M. *The Organization of Academic Work*, New York: John Wiley & Sons, 1973, p. 194.

TOOLKIT

## BOX 5-7    Stirring the Pot

Several institutions help researchers with similar interests cross departmental boundaries to respond to funding initiatives. Some have full-time staff associates (see Box 4-5); others rely on the vice provost for research. At the State University of New York (SUNY) at Stony Brook, Associate Vice President for Research Martin Schoonen brings teams together to respond to requests for applications (RFAs) and broad agency announcements.[a]

Schoonen's position is split 50:50 between a 3-year associate appointment as vice president for research and his position as professor in geosciences. The Office of the Vice President for Research maintains a Web page of annual program solicitations and distributes announcements for interesting talks on campus, high-profile papers, and the like. Says Shoonen: "I purposely do not organize seminars. I found that faculty are not looking for more talks to go to. They will come to a meeting if there is a funding opportunity."[b]

As a result of his matchmaking, at least two major projects have received funding. One was a US Agency for International Development award to help to rebuild Iraqi institutions of higher education.[c] The second was a National Institutes of Health award that brings together faculty interested in drug discovery and in tropical ecosystem conservation.[d]

For a pending NIH training grant with three other institutions and about 100 possible mentors, his office organized meetings to get potential mentors to sign on. They brought together a diverse group of faculty representing medical science, social science, environmental science, physical science, and economics. Virtually all paperwork associated with the grant application was handled by the Office of the Vice President for Research.

Efforts that have not led to awards have nevertheless been good investments in community-building. Some subsets of proposal-team members are working together on a small scale—for example, an economist with a nutritionist, a materials scientist with a microbiologist, and an environmental scientist with a virologist.

When an RFA calls for a multidisciplinary or interdisciplinary approach, he often "starts with calling some people I know. If there is some interest, I will convene a meeting. The meeting is announced campuswide. I have developed this strategy so that I know there will be some interested faculty (contacted directly by me) at the very least." However, through the campus announcement, he usually uncovers some additional people. For example, "our dean of libraries became a key player in the Iraq proposal. It turned out he had been trained in Near Eastern culture, can read Arabic, and had set up a library in Egypt." Once the team is formed, he guides them through the maze of proposal paperwork, reminds them of deadlines, helps organize meetings to work on the proposal, and creates an electronic home so that faculty can share files.

---

[a]Martin Schoonen, Associate Vice President for Research and Professor of Geochemistry, State University of New York, Stony Brook. Comments at the Convocation on Facilitating Interdisciplinary Research, Washington, DC. January 29, 2004.
[b]Martin Schoonen. Personal Communication. April 21, 2004.
[c]See *http://commcgi.cc.stonybrook.edu/artman/publish/article_573.shtml.*
[d]See *http://icte.bio.sunysb.edu/pages/ICBG_project.htm.*

## TOOLKIT

### BOX 5-8   Making Money Flow Sideways: Budgeting Models at UC Davis and the University of Michigan

**University of California, Davis**[a]

The proliferation of interdisciplinary programs in the 1970s challenged the old "vertical" funding model at universities with more "horizontal" programs that cut across college lines. "Money naturally runs downhill," writes Cristina Gonzalez, "and it is hard to make it flow sideways." UC Davis experimented with two ways to overcome this "law of gravity": distribute funds from a central office directly to interdisciplinary programs without going through the deans, and bring matching funds from a central office, such as the graduate school, to support the program.

UC Davis still does both, with increasing emphasis on matching funds. The Office of Graduate Studies (OGS) has the key role in supporting interdisciplinary programs, with an enrollment-based funding formula for administrative support of graduate groups. A few years ago, the formula was updated with a system of matching funds between the OGS and the college deans with the understanding that future matches by the college deans would come from their own budgets.

Gonzalez concluded that although the system works at UC Davis, universities have become too complex for a one-size-fits-all solution to the funding challenges of interdisciplinary programs. "Making money flow horizontally in a vertical funding system," she writes, "is highly customized engineering work that must take the individual characteristics of each campus into account."

**University of Michigan**[b]

The University of Michigan recently (FY 1998-1999) changed its budget model in ways more favorable to the management of interdisciplinary work, especially extradepartmental programs categorized as organized research units. In contrast with the previous system of "value-centered management," or incremental budgeting, the new budget system provides a mix of activity-based and discretionary budgeting. In activity-based budgeting, revenues flow preferentially toward units that are credited with larger revenue generation. At the same time, the revenue-generating activities generally create costs that must be covered. Through a balance of activity-based and discretionary budgeting, the provost and president retain considerable discretion in funding initiatives at the school, college, or research-unit level independently of current revenue-generating capacity. The system is designed to reserve flexible resources that can be reallocated across units.

---

[a]González, C. (2003) The Role of the Graduate School in Interdisciplinary Programs: The University of California, Davis Budget Model. CGS Communicator, Vol. XXXVI, Number 5.

[b]Courant, P. N. and Knepp, M. "Budgeting with the UB Model at the University of Michigan," Office of the Provost, University of Michigan, 2000. *www.umich.edu/~provost/budgeting/ubmodel.html.*

mon" and for new initiatives. Some institutions, including Columbia University, are using resources such as revenues generated from the licensing of intellectual property, to invest in new interdisciplinary research and teaching initiatives (see Box 5-8).

## CONCLUSIONS

It is possible for administrators of academic institutions to create supportive environments and policy structures that allow researchers to do their best—including interdisciplinary researchers, who face the special challenges summarized above.[16]

*"I understand they're going to connect them. The Provost ordered it."*

---

[16]See Holton, G., Chang, H., and Jurkowitz, E. "How a scientific discovery is made: A case history", *American Scientist*, Vol. 84, July-August 1996, pp. 364-75. The authors write that scientific innovation "depends on a mixture of basic and applied research, on interdisciplinary borrowing, on an unforced pace of work and on personal motivations that lie beyond the reach of the administrator's rule book" (p. 364).

Because research is difficult to manage, there are limits on the institution's ability to effect change; new fundamental knowledge cannot be produced on cue or on schedule. Nonetheless, the committee suggests that an institution can create an environment in which research flourishes by adapting organizational elements to its particular culture. Such an environment might be characterized by flexibility, a natural, unforced pace of work, and policies that promote borrowing and sharing within and between disciplines. As researchers find new collaborators, join new conversations, and enter new disciplinary cultures, they increase their opportunities to generate new understanding.

## FINDINGS

In attempting to balance the strengthening of disciplines and the pursuit of interdisciplinary research, education, and training, many institutions are impeded by traditions and policies that govern hiring, promotion, tenure, and resource allocation.

The success of IDR groups depends on institutional commitment and research leadership. Leaders with clear vision and effective communication and team-building skills can catalyze the integration of disciplines.

## RECOMMENDATIONS

### Academic Institutions' Policies

**I-1:  Academic institutions should develop new and strengthen existing policies and practices that lower or remove barriers to interdisciplinary research and scholarship, including developing joint programs with industry and government and nongovernment organizations.**

For example, institutions can

•   Provide more flexibility in promotion and tenure procedures, recognizing that the contributions of a person in IDR may need to be evaluated differently from those of a person in a single-discipline project. Institutions could

— Establish interdisciplinary review committees to evaluate faculty who are conducting IDR.

— Extend the venue for tenure review of interdisciplinary scholars beyond the department.

— Increase recognition of co-principal investigators' research activities during promotion and tenure decisions.

— Develop mechanisms to evaluate the contribution of each member of an IDR team.

• Establish institutional advisory committees of researchers successful in IDR to evaluate new proposals prior to implementation.

• Require regular reviews of IDR centers and institutes and establish sunset provisions, where appropriate, when they are initiated.

• Give high priority to recruitment of appropriate faculty and other researchers whose focus is interdisciplinary; this can be accomplished in part by allocating substantial resources to centrally funded, multidepartmental hiring of faculty and postdoctoral scholars and admission of graduate students.

• Coordinate hiring across departments and centers to maximize collaborative research and teaching possibilities.

• Develop joint IDR programs and internships with industry.

• Allow for the longer startup time required by some IDR programs.

• Gather information about the extent, quality, and importance of IDR in the institution and make the information available to faculty.

• Provide mechanisms to build a community of interdisciplinary scholars across the institution similar to the community that is in a department.

**I-2: Beyond the measures suggested in I-1, institutions should experiment with more innovative policies and structures to facilitate IDR, making appropriate use of lessons learned from the performance of IDR in industrial and national laboratories.**

For example, institutions can

• Experiment with alternatives to departmental tenure through new modes of employment, retention, and promotion.

• Selectively apply pooled faculty lines and funds available for startup costs for new faculty toward recruitment of faculty with interdisciplinary interests and credentials.

• Experiment with administrative structures that lower administrative and funding walls between departments and other kinds of academic units.

• Create laboratory facilities with reassignable spaces and equipment for people performing IDR.

• Create specific IDR grants and training programs for distinct career stages to assist in learning new disciplines and participating in IDR programs.

• Create mechanisms to fund graduate students and postdoctoral scholars whose research draws on multiple fields and may not be considered central to any one department.

- Develop a process for dealing with intellectual-property allocation that is consistent with encouraging IDR.
- Increase "porosity" across organizational boundaries by
   — Encouraging joint recruitment and appointment of faculty through resources available centrally.
   — Creating opportunities for faculty to compete for internal leave for study in a new discipline so as to take courses, training, and additional advanced degrees in their own universities.
   — Encouraging departments and colleges to work with IDR centers and institutes in hiring faculty with interdisciplinary backgrounds.
   — Providing fellowships that are portable within the institution.
   — Allowing courtesy appointments that recognize interactions and collaborations across departments but that do not have the formal split responsibility of a joint appointment.
   — Placing departments near one another to take advantage of their potential for fruitful interdisciplinary collaborations.

**I-3: Institutions should support interdisciplinary education and training for students, postdoctoral scholars, researchers, and faculty by providing such mechanisms as undergraduate research opportunities, faculty team-teaching credit, and IDR management training.**

Such education and training could cover interdisciplinary research techniques, interdisciplinary team management skills, methods for teaching non-majors, etc. For example, institutions can

- Provide more opportunities for undergraduate interdisciplinary research experiences.
- Allow faculty to receive full credit for team teaching in interdisciplinary courses.
- Encourage multiple mentors for students and pairing of appropriate senior interdisciplinary faculty with junior ones interested in IDR.
- Provide opportunities (such as sabbaticals) for students and faculty members to learn the content, languages, and cultures of disciplines other than their own, both within and outside their home institution.
- Support formal programs on the management of IDR programs, including leadership and team-forming activities.

**I-4: Institutions should develop equitable and flexible budgetary and cost-sharing policies that support IDR.**

For example, institutions can

- Streamline fair and equitable budgeting procedures across department or school lines to allocate resources to interdisciplinary units outside the departments or schools.

• Create a campuswide inventory of equipment to enhance sharing and underwrite centralized equipment and instrument facilities for use by IDR projects and by multiple disciplines.

• Credit a percentage of all projects' indirect costs to support the infrastructure of research activities that cross departmental and school boundaries.

• Allocate research space to projects, as well as departments.

• Deploy a substantial fraction of flexible resources—such as seed money, support staff, and space—in support of IDR.

## Team Leaders

**T-1: To facilitate the work of an IDR team, its leaders should bring together potential research collaborators early in the process and work toward agreement on key issues.**

For example, team leaders can

• Catalyze the skillful design of research plans and the integration of knowledge and skills in multiple disciplines rather than "stapling together" similar or overlapping proposals.

• Establish early agreements on research methods, goals and timelines, and regular meetings.

**T-2: IDR leaders should seek to ensure that each participant strikes an appropriate balance between leading and following and between contributing to and benefiting from the efforts of the team.**

For example, leaders can

• Help the team to decide who will take responsibility for each portion of the research plan.

• Encourage participants to develop appropriate ways to share credit, including authorship credit, for the achievements of the team.

• Acquaint students with literature on integration and collaboration.

• Provide adequate time for mutual learning.

# 6

# How Funding Organizations Can Facilitate Interdisciplinary Research

Many kinds of organizations in addition to academic institutions provide funding for scientific and engineering research, including federal and state agencies, private foundations, corporations, and nonprofit organizations. Congress and state legislatures also play major roles in determining research priorities for the nation and states. All funding organizations, because of the financial and other resources they can potentially bring to bear, can develop and press for reforms that facilitate interdisciplinary research and education.

## A VISION FOR FUNDING ORGANIZATIONS THAT WISH TO PROMOTE INTERDISCIPLINARY RESEARCH

Some organizations may hesitate to become involved in interdisciplinary research (IDR) to the extent that it requires risk-taking and administrative complexities that may be greater than those of single-discipline programs. It is helpful, however, to recall the "drivers" of IDR described in Chapter 2, which indicate that today's most pressing research and societal questions are often best addressed by interdisciplinary approaches.

> **Convocation Quote**
>
> In our mind, it takes a multidisciplinary approach to address a number of critical, important problems that are mission areas for the Defense Department, but we also think that it does three things. It accelerates research progress by bringing groups of people together to address the problem. It expedites the transition of research into products that can actually be used by the Defense Department and the community in general. Most importantly perhaps, it prepares students to think in an interdisciplinary manner and prepares them to be a more agile sort of workforce.
>
> William Berry, director, Office of Basic Research, Department of Defense

Those whose missions are aligned with IDR can promote it with several strategies. They may wish to have a substantial influence on the direction and productivity of research, support emerging fields that have insufficient support elsewhere, emphasize the educational and training components of interdisciplinary work, or develop more effective evaluation and review measures that help to select and sustain the best projects and people. By pulling and adjusting their own levers of influence, funding organizations play a critical role in facilitating IDR.

Congress has shown its support of IDR, indicating, for example, in the fiscal year 2004 Consolidated Appropriations Act that the National Science Foundation's Research and Related Activities account "supports . . . critical cross-cutting research which brings together multiple disciplines. The conferees urge the Foundation in allocating the scarce resources provided in this bill and in preparing its fiscal 2005 budget request to be sensitive to maintaining the proper balance between the goal of stimulating interdisciplinary research and the need to maintain robust single issue research in the core disciplines."[1]

## BARRIERS ENCOUNTERED BY FUNDING ORGANIZATIONS IN SUPPORTING IDR

Like academic institutions, funding organizations may face significant barriers in facilitating IDR—some that originate in their own traditions and others that are inherent in the nature of IDR. Most of the barriers discovered during this study have to do with the complex nature of IDR:

---

[1]US Congress. House Committee on Appropriations. 2003. H.R. 2673—*Making Appropriations for Agriculture, Rural Development, Food and Drug Administration, and Related Agencies for the fiscal year ending September 30, 2004, and for other purposes.* 108th Cong., H.R. 108-401:1167.

- Effective review of IDR proposals may not be possible with traditional peer review that relies primarily on experts in a single discipline.
- Funding organizations often find IDR programs more difficult to plan and develop than single-discipline programs because IDR programs may require extra time to build consensus and introduce researchers to new languages, knowledge, and cultures.
- Funders, like other organizations, have insufficient knowledge about the best ways to solicit IDR proposals and evaluate IDR programs.
- It is not always easy to manage the transition of an IDR project from startup to larger-scale, longer-term project funding so as to maintain program momentum.

Among the top recommendations to funding agencies from survey respondents were developing strategies to facilitate IDR, implementing a more effective review process, and rethinking funding allocation strategies (see Figure 6-1).

Barriers to IDR exist even in the most experienced funding organizations. For example, the National Science Foundation (NSF) has long been a pioneer in promoting IDR (see Figure 6-2). A recent comprehensive study of NSF policies found that "NSF's priority areas demonstrate an interdisciplinary perspective, especially as evidenced by the extent of cross-directorate funding." However, the same study cautioned that "highly variable attention to interdisciplinary research in NSF's strategy, budget, and public documents does not communicate a consistent message." It also found that "no effective mechanism is in place to track or set performance goals for interdisciplinary research that can be used for planning, budget, and management decision making" and that "NSF's two merit review criteria say relatively little with regard to interdisciplinary research."[2]

Additional difficulties were reported by investigators applying for agency funding. Applicants interested in interdisciplinary work felt disadvantaged relative to applicants focusing on single disciplines because of relatively short submission deadlines, pressure to understate costs for IDR proposals, the page limit on proposals, the difficulty of teaming administratively with investigators in different institutions, and lack of a well-defined review path for IDR proposals.[3]

Those findings may say more about the general challenges of funding any large IDR project or initiative than about the shortcomings of a particular agency. Indeed, state and federal budgeting systems, combined with

---

[2]National Academy of Public Administration (NAPA), "National Science Foundation: Governance and Management for the Future," April 2004, pp. 61-89.

[3]NAPA, ibid. p. 102.

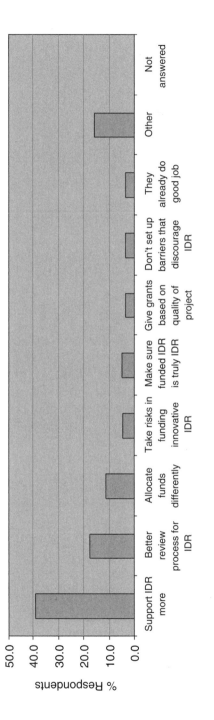

**FIGURE 6-1** Recommendations to funding agencies.

NOTES: Survey Question: "If you could recommend one action that funding agencies could take that would best facilitate interdisciplinary research, what action would that be?" The top three recommendations for funding agencies (n = 266) were to provide more support for IDR (39.1 percent), to develop and implement a more effective review process for IDR proposals (17.7 percent), and to rethink funding allocation strategies (11.3 percent).

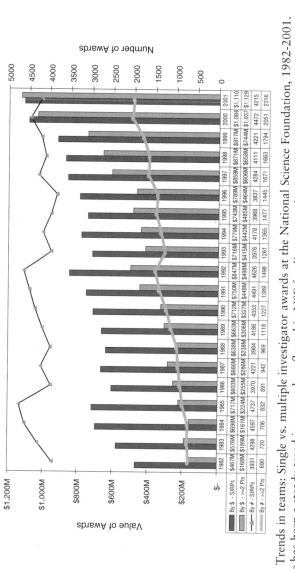

**FIGURE 6-2** Trends in teams: Single *vs.* multiple investigator awards at the National Science Foundation, 1982-2001.

| | 1982 | 1983 | 1984 | 1985 | 1986 | 1987 | 1988 | 1989 | 1990 | 1991 | 1992 | 1993 | 1994 | 1995 | 1996 | 1997 | 1998 | 1999 | 2000 | 2001 |
|---|---|---|---|---|---|---|---|---|---|---|---|---|---|---|---|---|---|---|---|---|
| By $ - SIRPs | $467M | $576M | $698M | $717M | $603M | $666M | $638M | $663M | $712M | $750M | $847M | $716M | $779M | $743M | $789M | $859M | $871M | $917M | $1,084 | $1,110 |
| By $ - >=2 PIs | $168M | $189M | $161M | $224M | $255M | $286M | $238M | $306M | $327M | $448M | $498M | $415M | $442M | $485M | $464M | $606M | $658M | $744M | $1,037 | $1,129 |
| By # - SIRPs | 3931 | 4298 | 4597 | 4737 | 3970 | 4221 | 3984 | 4186 | 4353 | 4491 | 4626 | 3976 | 4178 | 3968 | 3837 | 4284 | 4111 | 4221 | 4472 | 4215 |
| By # - >=2 PIs | 690 | 720 | 706 | 832 | 891 | 942 | 969 | 1118 | 1227 | 1389 | 1498 | 1261 | 1365 | 1477 | 1445 | 1671 | 1663 | 1794 | 2051 | 2016 |

NOTES: There has been a steady trend in research, reflected in NSF funding, towards awards with more than one PI. This predates any recent initiative and actually helped influence the move toward interdisciplinary initiatives in recent years. While the growth area is in multiple PI awards, single investigators won about half of the funding and two-thirds of the awards in fiscal year 2001. This compares with seven-eighths of the funding and three-quarters of the grants for single investigator research proposals in fiscal year 1982.

SOURCE: From Vernon Ross, Budget Operations and Systems, NSF, from a presentation given at the National Academy of Engineering, October 3, 2003. While the NSF supports IDR, it has no operational definition of IDR. In a study of NSF which covered IDR, the National Academy of Public Administration concluded that the best current measure of IDR at NSF is multi-investigator grants. (See footnote 2.)

long-term mission strategies, place substantial constraints on the funding activities of every public agency.

---

**Convocation Quote**

No single program can be a forcing function. Creativity, good research, good ideas and research questions are not owned by a single program. Dedicated program champions at both DOE headquarters and at the laboratories are critical because integration needs leadership.

Marvin Singer, senior adviser, Applied Energy Program,
Office of Science, Department of Energy

---

## SUPPORT FOR IDEAS AND INITIATIVES

Nonetheless, some funding organizations, especially federal agencies with research-based missions, have built large IDR programs by responding to the drivers of IDR described in Chapter 2, especially the inherent complexity of nature, the desire to follow questions to the interfaces of disciplines, and the need to address multifaceted societal issues.

Both public and private funding organizations have been successful in linking their missions with an interdisciplinary vision. NSF, the only agency whose primary mission is to support science and engineering research and education, has been a leader and exemplar in supporting individuals, projects, and multi-institution programs for IDR. Its science and technology centers and engineering research centers, for example, have served as a model for interdisciplinary centers at universities that work in partnership with industry (see Box 8-2) and its research training grants (see Box 8-4). Other IDR initiatives include the Mathematical Sciences: Innovations at the Interface, the Biocomplexity in the Environment: Integrated Research and Education in Environment Systems program, and the former Information Technology Research program.

Similarly, the National Institutes of Health (NIH) has adopted its own interdisciplinary vision. Noting that "the traditional divisions within biomedical research may in some instances impede the pace of scientific discovery,"[4] NIH has constructed a new strategic roadmap intended "to lower these artificial organizational barriers and advance science." To do so, the agency has announced a series of awards specially aimed at supporting IDR (see Box 6-1), including awards for "training of scientists in interdiscipli-

---

[4]*http://nihroadmap.nih.gov/interdisciplinary/index.asp.*

---

**INNOVATIVE PRACTICE**

**BOX 6-1   NIH Roadmap: Research Teams of the Future**

Interdisciplinary research is an important initiative of the 2003 National Institute of Health (NIH) Roadmap.[a] One of the three themes of the roadmap is "Research Teams of the Future." It is becoming more obvious that as research problems become more complex it is often necessary to amalgamate a research team with many disciplines to tackle a research problem effectively. However, NIH found that the traditional divisions in biomedical research in some instances may impede scientific discovery. The purpose of their IDR initiative is to develop innovative ways to combine skills and disciplines to accelerate discovery of fundamental knowledge and advance existing knowledge.

Several grants and funding opportunities were created to help to facilitate IDR.[b] Included are training grants for graduate students and postdoctoral scholars (the T90),[c] a curriculum development award,[d] and a short intensive course for researchers at all career levels to receive formal training in another discipline.[e] The goal of the various programs is for researchers to "emerge with sufficient understanding of a new discipline(s) that they can meld it with their previous training to generate new interdisciplines with novel research strategies."

---

[a]Roadmap home page *http://nihroadmap.nih.gov/index.asp.*

[b]NIH Roadmap Interdisciplinary Initiative home page *http://nihroadmap.nih.gov/interdisciplinary/grants.asp.*

[c]Kozel, P. "NIH's Roadmap to the Future" *Science's Next Wave.* January 2004. *http://nextwave.sciencemag.org/cgi/content/full/2004/01/08/4?* Training for a New Interdisciplinary Research Workforce (T90) *http://grants.nih.gov/grants/funding/t90.htm.*

[d]Curriculum Development Award in Interdisciplinary Research (RFA-RM-04-007) *http://grants.nih.gov/grants/guide/rfa-files/RFA-RM-04-007.html.*

[e]Short Programs for Interdisciplinary Research Training (RFA-RM-04-008) *http://grants.nih.gov/grants/guide/rfa-files/RFA-RM-04-008.html.*

---

nary strategies; creation of specialized centers to help scientists forge new and more advanced disciplines from existing ones; and initiation of forward-looking conferences to catalyze collaboration among the life and physical sciences, important areas of research that historically have had limited interaction."

Other major federal efforts are explicitly interdisciplinary in concept, including the Multidisciplinary Research Program of the University Research Initiative (MURI), a multi-agency Department of Defense program that supports research teams "whose efforts intersect more than one traditional science and engineering discipline"[5] (see Box 6-2); the Interagency

---

[5]*http://www.onr.navy.mil/sci_tech/industrial/muri.htm.*

---

**INNOVATIVE PRACTICE**

**BOX 6-2　The Department of Defense Multidisciplinary University Research Initiative**

One way for funding organizations to support interdisciplinary research is to establish specific grants programs that reward interdisciplinary approaches. The US Department of Defense (DoD) Multidisciplinary University Research Initiative (MURI)[a] is specifically targeted at proposals that "intersect more than one traditional science and engineering discipline." The program disperses about $150 million per year and represents about 10 percent of DoD's overall basic research program.

MURI is designed to complement the core research supported by the department, which consists primarily of single-investigator approaches. Examples of subjects to be considered for funding by MURI are "Hybrid Bio-Mechanical Systems" and "Micro Hovering Aerial Vehicles with an Invertebrate Vision Inspired Navigation System." Goals for the program include bringing researchers together to expedite discovery and training students to think in an interdisciplinary manner. Like many examples of IDR, these grants are motivated by specific engineering goals that require advances in basic understanding to occur at the interface of diverse fields. As William Berry, the director of the Office of Basic Research at the Pentagon, has said, "Think of the end at the beginning."[b] In this case, the engineering goals are derived partially from the mission of the funding agency itself and partially from the researchers' vision for the kind of technology they want to create.

---

[a]MURI 2004 Program Solicitation. *http://www.onr.navy.mil/sci%5Ftech/industrial/363/muri.asp.*
[b]Berry, B. Comments at Convocation on Facilitating Interdisciplinary Research. January 29, 2004, Washington, D.C. *http://www7.nationalacademies.org/interdisciplinary/Convocation_Agenda.html.*

---

Education Research Initiative (IERI);[6] the National Aeronautics and Space Administration (NASA) astrobiology program (see Box 6-3);[7] and the government-wide National Nanotechnology Initiative, beginning in FY 2005, of which NSF will have the largest share.

---

[6]The IERI pairs information-technology (IT) researchers with those in another field interested in using cutting-edge IT to help solve problems. The goal of the initiative is to support scientific research that investigates the effectiveness of educational interventions in reading, mathematics, and the sciences as they are implemented in varied school settings with diverse student populations. *http://www.ed.gov/offices/OERI/IERI/.*

[7]For this program, according to NASA, "interdisciplinary research is needed that combines molecular biology, ecology, planetary science, astronomy, information science, space exploration technologies, and related disciplines. The broad interdisciplinary character of astrobiology compels us to strive for the most comprehensive and inclusive understanding of biological, planetary and cosmic phenomena." *http://astrobiology.arc.nasa.gov/.*

INNOVATIVE PRACTICE

**BOX 6-3    NASA Fosters the Development of
Interdisciplinary Fields**

Federal agencies can play a pivotal role in launching IDR by providing funding for developing fields. As part of its Origins Program,[a] the National Aeronautics and Space Administration (NASA) has committed to promoting research in astrobiology, the interdisciplinary study of life in the universe.

The NASA Astrobiology Institute (NAI)[b] was created in 1998. Initially, 11 research proposals were selected; today, there are 16 participating institutions. Lead teams are supported by NASA through 5-year cooperative agreements with the Ames Research Center. Team members are from different disciplines—including physics, astronomy, geology, and biology—and often from different geographic locations. A major goal of NAI is to train a new generation of astrobiologists; to this end, NAI sponsors seminars, workshops, and professional training courses.

In addition to the lead teams, NAI fosters astrobiology research through support of research focus groups. These groups typically stimulate new fields of research and promote collaborations within and outside of NAI. Focus group proposals are typically for 3 years. NAI provides support for postdoctoral fellowships through the NASA-National Research Council Associateship Program. An NAI research scholarship provides stipends and research-related travel funds to graduate students and postdoctoral scholars so that they can circulate between two or more of the lead teams.

---

[a]NASA Origins Program home page *http://origins.jpl.nasa.gov/index1.html.* Accessed April 30, 2004.

[b]Astrobiology Institute home page *http://nai.arc.nasa.gov/index.cfm.* Accessed April 30, 2004.

The Defense Advanced Research Projects Agency (DARPA) is a subagency that has served as a global model for interdisciplinary effectiveness (see Box 6-4). By promoting high organizational flexibility and lowering barriers to collaboration, DARPA has been able to support innovative, cross-disciplinary projects at every level of complexity, including the open-ended research that led to major features of the Internet. Its Defense Science Office draws program officers from diverse disciplines and has directed strong support toward IDR projects. The R&D structure of the Department of Homeland Security was modeled explicitly on DARPA, as recommended by the National Academies.[8]

An essential feature of such new funding models is innovative, risk-taking leadership in the funding body. For example, those mentioned above

---

[8]National Research Council. *Making the Nation Safer: The Role of Science and Technology in Countering Terrorism.* 2002. Washington, DC: The National Academies Press, pp. 335-57.

EVOLUTION

## BOX 6-4    The Defense Advanced Research Projects Agency

The Defense Advanced Research Projects Agency (DARPA),[a] created in 1957 in the wake of Sputnik, has a long record of supporting high-risk, interdisciplinary research. DARPA is probably best known for its support of the ARPANET, the precursor to today's Internet, and stealth technology. In 1960, it began to fund the interdisciplinary laboratories, which played a critical role in fostering materials science and engineering in the United States. By the time DARPA transferred the program to the National Science Foundation in the early 1970s, it was supporting 600 faculty in physics, chemistry, metallurgy, materials science and engineering, and electrical engineering. More recently, DARPA launched a research program in FY 2000 called Bio:Info:Micro,[b] which funded six interdisciplinary teams of researchers in biology, information technology, and microsystems technology to deepen our understanding of neuroprocessing and regulatory networks.

DARPA has been successful in supporting high-risk, high-return IDR for a number of reasons,[c] among them:

1. Solicitations are focused on hard problems or emerging scientific and technical opportunities, not disciplines.
2. Offices are not organized around disciplines. At least 13 science, engineering, and medical disciplines are represented in the 20-person technical staff of DARPA's Defense Science Office.
3. The Department of Defense is willing to invest a small percentage of its budget (less than 1 percent) in radical innovation, but this tiny fraction of their budget is substantial—$3 billion.
4. DARPA continuously recruits high-quality program managers, who generally stay for 4-6 years. This ensures a steady stream of new ideas.
5. DARPA program managers are responsible for developing research programs. They define the problems, typically through continuous interactions with the research community on the one hand and the user community on the other hand. Thus, they are familiar both with the national technology capabilities that need to be developed and with the cutting-edge science and engineering issues, barriers, and opportunities that, if addressed with serious resources and creative interdisciplinary approaches, might lead to revolutionary advances.
6. DARPA program managers not only develop the programs but manage proposal solicitation and selection. Thus, they have complete control over which proposals to fund. They encourage risky and less mature ideas than are normally tolerated at agencies that rely on the more traditional peer-review process.
7. DARPA has no "entitled constituencies" and can fund research in academe, industry, and national laboratories
8. DARPA is willing to fund larger grants, which are often necessary to put together a "critical mass" of researchers in different disciplines.
9. DARPA program managers often play a hands-on role in encouraging interaction between the research teams they are funding.

---

[a]DARPA home page. *http://www.darpa.mil/*. Accessed April 30, 2004.

[b]Bio:Info:Micro Program Solicitation. *http://www.darpa.mil/baa/ra00-14.htm*.

[c]Dubois, L. H. "DARPA's Approach to Innovation and Its Reflection in Industry." In *Reducing the Time from Basic Research to Innovation in the Chemical Sciences*. A Workshop Report of the Chemical Sciences Roundtable. 2003. Washington, D.C.: The National Academies Press.

have all encouraged exploratory and frontier research beyond the perceived boundaries of disciplines by providing opportunities for networking and special initiatives. Some funding organizations have also developed new proposal-review procedures to ensure expertise in each discipline represented in a project or proposal.

Finally, funding organizations can promote more public-private collaboration. In Europe, the Organization for Economic Co-operation and Development (OECD) Futures Projects offer a pragmatic approach to focused, multidisciplinary research and policy analysis on future-oriented themes involving both governments and private-sector participants. Futures Projects are launched when there is no appropriate committee or directorate to address a theme or when the interdisciplinary nature of the theme does not lend itself easily to treatment by a single or even several directorates.[9]

## SUPPORT FOR PEOPLE AND PROGRAMS

In addition to funding new ideas and initiatives, funding organizations can focus their resources on opportunities to fund programs and people at various stages of their careers and in curriculum reform and interdisciplinary education. The stages, described in Chapter 4, have considerable overlap in the sense that all researchers, from undergraduates to senior faculty, have interests and motivations in common and benefit from similar kinds of support in addressing interdisciplinary research and education.

**Convocation Quote**

When MacArthur selects people to participate in research networks, it is more about their interest to go beyond their own paradigm and to be interested in a collaborative endeavor. Leadership is the key in terms of the success of our network. These people are honest brokers. They are generative. They are intellectually curious. They are about facilitating the work.

Laurie Garduque, program director for research at the John D. and Catherine T. MacArthur Foundation

---

[9]OECD Futures Program home page. *http://www.oecd.org/department/0,2688,en_2649_33707_1_1_1_1_1,00.html*. Accessed April 30, 2004.

## Graduate Students

A goal of the National Science Foundation's (NSF) Integrative Graduate Education and Research Traineeships (IGERT) program (see Boxes 4-1 and 8-4) is to "prepare scientists for careers at the interstices of disciplines and in non-traditional settings."[10] The IGERT program has particular relevance to this study in that it was stimulated in part by a previous Committee on Science, Engineering, and Public Policy (COSEPUP) report on graduate education.[11] The training grants, which are allocated to institutions and then to the students themselves, are especially important in light of reduced support for graduate students by some agencies and foundations.[12]

## Postdoctoral Scholars

Funding organizations can consider shifting some of their resources to supporting postdoctoral scholars. Postdoctoral scholars with a solid base in one discipline may become more productive if they have opportunities to learn and work in additional disciplines. Such support can be used for additional training, laboratory visits, and coursework. The Burroughs Wellcome Foundation supports an Interfaces in Science program that provides transitional funding for postdoctoral scholars and faculty with backgrounds in physics, mathematics, computer science, and engineering who want to explore aspects of biology[13] (see Box 6-5).

## Faculty

A useful mechanism for junior or senior faculty to gain new skills and master new disciplines is a portable fellowship, such as that in the Sloan Fellows Program,[14] which can be designed for use in the institution or beyond. Such support may be hard to find in the traditional salary or grant

---

[10]Hackett, E. J. "Initiatives at the U.S. National Science Foundation," In Weingart, P. and Stehr, N. *Practising Interdisciplinarity*. Toronto: Unviersity of Toronto Press. 2000, p. 251. The NSF IGERT program states "the program is intended to catalyze a cultural change in graduate education, for students, faculty, and institutions, by establishing innovative new models for graduate education and training in a fertile environment for collaborative research that transcends traditional disciplinary boundaries." *http://www.nsf.gov/pubs/2004/nsf04550/nsf04550.htm.*

[11]National Research Council, *Reshaping the Graduate Education of Scientists and Engineers*, Washington, DC: National Academy Press. 1995.

[12]For example, the Howard Hughes Medical Institute recently ended its research training fellowships for graduate students.

[13]*http://www.bwfund.org/programs/interfaces/index.html.*

[14]*http://www.sloan.org/programs/scitech_fellowships.shtml.*

**INNOVATIVE PRACTICE**

**BOX 6-5   Burroughs Wellcome Fund Career Transition Awards**

In 2002, the Burroughs Wellcome Fund (BWF) established a grant program to support young investigators working at the interfaces between biology and other disciplines. The program, titled Career Awards at the Scientific Interface, recognizes the potential role that physical, chemical, and computational sciences can play in innovative biological fields, such as genomics, quantitative structural biology, systems modeling, and nanotechnology.[a] In 2002, eight postdoctoral students were awarded grants; in 2003, seven grants were awarded.

Like Burroughs Wellcome's original career-awards program,[b] which was designed to facilitate the critical transition from postdoctoral training to tenure-track faculty positions, the Scientific Interface program provides $500,000 over 5 years to support 2 years of advanced postdoctoral training and 3 years of a faculty appointment. The program specifically encourages interdisciplinary work and training. First, candidates are required to hold a PhD in chemistry, physics, mathematics, computer science, statistics, or engineering and must propose a research project that addresses questions in biomedical science. Second, the foundation expects award recipients to continue their interdisciplinary cross-training and provides grant funds for travel to scientific meetings and for advanced coursework in biology. Finally, award recipients are required to form collaborations with well-established investigators outside their own fields.

---

[a]Burroughs Wellcome Fund. *2005 Career Awards at the Scientific Interface. http://www. bwfund.org/programs/interfaces.*

[b]Pion, G. and Ionescu-Pioggia, M. "Bridging Postdoctoral Training and a Faculty Position: Initial Outcomes of the Burroughs Wellcome Fund Career Awards in the Biomedical Sciences." *Academic Medicine.* 2003, 78(2):177-186.

structure. A more flexible option may be to support summer immersion experiences (see Box 4-1) or grants for workshops in emerging areas (see Box 6-3).

Similarly, funding organizations can spur fledgling IDR initiatives by providing seed money. Like venture funding at the early stage of formation of a firm, seed funding provides flexibility that is not available in many grants to shape innovative or experimental programs. Even modest amounts of seed funding can have a strong catalytic value in supporting demonstrations and visible pilot programs (see Box 6-6). Following that strategy, the Mellon Foundation provides some flexible funding for junior faculty engaged in IDR; similarly, the Beckman Foundation issued a request several years ago for proposals for high-risk IDR deemed insufficiently developed for funding by large agencies. Initiatives of those kinds are often more appropriate for private foundations than for federal agencies, which tend to fund programs that have already been launched.

---

EVOLUTION

## BOX 6-6    Fullerene Research at Rice University

In 1993, a faculty task force led by Richard Smalley defined a nanotechnology initiative at Rice University that built on interdisciplinary strengths in science and engineering. By 1997, several new faculty members had been hired, a new 70,000-ft$^2$ laboratory had been completed, and the Center for Nanoscale Science and Technology opened its doors.[a,b] The interdisciplinary research infrastructure provided by the center has provided Rice University a leadership role throughout the transition from basic research to development and commercialization of nanotube technologies.

The wide diversity of scientific applications for fullerene-based molecules not only laid the foundation for extensive interdisciplinary collaboration among scientists at Rice but helped to foster worldwide interest in carbon compounds. The carbon technology has made it possible to produce superconducting salts, three-dimensional polymers, catalysts, materials with new electric and optical properties, sensors, nanotubes,[c] and solar cells.[d]

Grants from the National Science Foundation, the National Aeronautics and Space Administration (NASA), and the Department of Defense have funded the development of laser-oven production facilities, which became the commercial operation Tubes@Rice Inc. for supplying the world with research quantities of nanotubes. That process was licensed to DuPont for its use in manufactured display technologies, and DuPont and NASA purchased Rice's laser-oven single-wall nanotube (SWNT)-generating apparatus. A more scalable process based on a conversion of carbon monoxide to SWNTs was then developed. Called the HiPco process (high partial pressure of carbon monoxide), it was patented and commercialized by Rice University.[e]

---

[a]Center for Nanoscale Science and Technology Web page *http://cnst.rice.edu/index.cfm.*

[b]In the midst of this campaign, Rice University chemists Smalley and Curl with colleague Harold Kroto were awarded the Nobel Prize for their unique work with buckminsterfullerene, clusters of 60 carbon atoms (C60) that are bound into a stable and symmetric soccer ball configuration The Royal Swedish Academy of Sciences. *Press Release: The 1996 Nobel Prize in Chemistry. http://www.nobel.se/chemistry/laureates/1996/press.html.*

[c]Shelley, S. *Carbon Nanotubes: A Small-Scale Wonder. Chemical Engineering,* February 2003.

[d]Bethune, D. S. and Johnson, R. D. "Atoms in carbon cages: The structure and properties of endohedral fullerenes." *Nature* 366:123-29.

[e]Nanotechnologies Inc. Web site: *http://www.cnanotech.com/.* Accessed March 29, 2004.

---

## SUPPORT FOR INSTITUTIONS AND FACILITIES

A third strategy that funding organizations can follow is to support new institutions or facilities or to provide support to existing institutions for reforms or innovations that cannot be achieved under current conditions. For example, some funding organizations have chosen to support major new programs and centers by providing essential space, specialized personnel, and facilities:

• The University of Illinois at Urbana-Champaign negotiated a $40 million gift from the Arnold and Mabel Beckman Institute to build an interdisciplinary research center. The initiative began when the vice-chancellor asked faculty to develop an IDR proposal (see Box 5-6).
• Stanford University negotiated a gift from Jim Clark, founder of Netscape, to build the Bio-X facility, explicitly designed to foster IDR in biology and medicine. Research proposals and decisions about which researchers will receive space at the new facility are faculty-initiated. The facility brings together biologists, clinicians, engineers, chemists, physicists, and computer scientists to stimulate innovative thinking (see Box 9-6). Janelia Farm, conceived and funded by the Howard Hughes Medical Institute, is a similar building designed expressly to foster IDR (see Box 6-7).
• The Fred Kavli Foundation was recently formed to support three interdisciplinary fields: cosmology, neuroscience, and nanoscience. The foundation has funded nine institutes in universities (eight in the United States and one in Europe), has created four professorships at California universities and will begin awarding research prizes in 2007.

Specific funding and support mechanisms may help institutions facilitate IDR:

• *Encourage proposals that have multiple Principal Investigators (PIs).* They can supplement the standard model of funding a single investigator by funding IDR teams. Grants inviting team proposals can provide explicit recognition of the effectiveness of collaborative leadership.
• *Fund the collaborative process as well as interdisciplinary team research.* Rather than focusing funding wholly on research, funding organizations can experiment with funding the collaborative process, which includes travel, meetings, training, and other activities through which investigators learn one another's language, culture, and knowledge. In the committee's survey, respondents' top recommendations for institutions, project leaders, principal investigators, educators, postdoctoral scholars, and students focused on enhancing communication between researchers. Over 20 percent of the respondents stated specifically that interdisciplinary researchers need time to develop effective networks and research strategies.

INNOVATIVE PRACTICE

## BOX 6-7 Creating Spaces for Interdisciplinary Research[a]

Slated for completion in early 2006, the Howard Hughes Medical Institute (HHMI) Janelia Farm Research Campus will serve as an intellectual hub for several hundred scientists in diverse disciplines. HHMI expects to spend about $500 million to construct the campus and put its scientific programs into place. The initial construction will provide the laboratories to accommodate a permanent research staff of 200-300. Additional laboratories and facilities will be built for visiting researchers and for core scientific support staff and administration. Janelia Farm includes about 760,000 ft$^2$ of space, housing the research laboratories and support areas, a conference center, and housing for more than 100 visitors.

The scientific programs at Janelia Farm are designed to further collaboration and flexibility among scientists. Research teams will be kept small, and team leaders are expected to stay actively involved in bench research, not just manage or guide it.

Janelia Farm's two primary scientific agendas are to establish a continuing research program at the interface of emerging technologies and their application to biomedical problems and to make available project-oriented "surge" space where visitors can come together and use new technologies to solve problems. Janelia Farm provides the facilities, finances, and freedom for scientists to pursue collaborative, interdisciplinary research, bringing members of their research groups, to work for periods ranging from a few weeks to several years.

The architectural designs of the buildings and the laboratories are aimed at achieving both of Janelia Farm's central objectives—collaboration and flexibility. Thus, design is guided by three principles that HHMI has gleaned from its considerable experience in creating successful work environments for scientists:

- Understand the researchers' needs and their preferences.
- Keep work spaces standardized and rational.
- Make the spaces adaptable to accommodate changes in research.

---

[a]Janelia Farm home page *http://www.hhmi.org/janelia/*. Accessed April 30, 2004.

## Convocation Quote

The calls we got from grantees in our interdisciplinary science program were not about extensions to the grants or budget. They were, "Could you help us figure out how to get the collaboration to work more effectively?" Collaboration is the bedrock of interdisciplinary research work. That is an area we think a funder interested in fostering interdisciplinary work ought to focus on: "glue money" to support meetings, bringing people together, travel, learning how to work together, and some of the team training aspects.

Barry Gold, Program Officer, Conservation and Science,
The David and Lucile Packard Foundation

- *Make grants of longer duration.* Longer-term grants, with sufficient safeguards to ensure that progress is being made, can be helpful in supporting IDR efforts because extended startup periods are often required.
- *Fund studies of the social aspects of the interdisciplinary process.* There is insufficient understanding of the motivations, modes of working, external pressures from the larger community, and other aspects of initiating and sustaining IDR in a given environment. A valuable contribution would be funding for research on the creation and implementation of new models for providing the interactions and dialogues that hold IDR together, such as "collaboratories."

Recent interagency discussions and focus groups with researchers and university administrators sponsored by the US Office of Science and Technology Policy found several areas of agreement on how to facilitate IDR (see Box 6-8). Many of their findings parallel and support those in the present report.

## REVIEWING PROPOSALS FOR INTERDISCIPLINARY ACTIVITIES

Funding organizations, through the mechanisms they use to approve or reject grant proposals, have great influence over the kinds of research proposals that are funded in this country. As discussed in Chapters 5 and 8, those mechanisms often evaluate proposals from the point of view of one or several disciplines, by using review panels that may have little expertise in IDR. Expertise in IDR, as well as in the constituent and related disciplines, is required to review multidisciplinary projects fairly and award credit for the contributions of project members.

Funding organizations can help to improve the review process in at least two ways. First, they can reform their own mechanisms of review by ensuring adequate breadth among the pool of researchers who review IDR proposals, in addition to the necessary depth of expertise in specific disciplines. Second, they can support additional study and experimentation with current and alternative mechanisms for reviewing IDR.[15] Funders might consider, as an example, the multistage process familiar in Europe, where the judgment of disciplinary experts is combined in various ways with the

---

[15]In its recent study of NSF funding procedures, the National Association of Public Administration recommended that "NSF ensure that review procedures for interdisciplinary research are transparent" and "NSF establish supplementary review criteria that will help to assess the quality of interdisciplinary effort in those programs where both single and multiple discipline proposals compete for a common pool of funds." NAPA, *National Science Foundation: Governance and Management for the Future*, April 2004.

TOOLKIT

## BOX 6-8 OSTP Business Models Initiative

In spring 2003, the National Science and Technology Council of the Office of Science and Technology Policy established the Research Business Model subcommittee to find out more about the changing nature of scientific research and how the changes are affecting the success of research sponsored by federal agencies. Through a series of workshops,[a] the subcommittee learned more about how research is being performed and how federal agencies might improve support of research that is interdisciplinary.[b]

Working groups identified two main drivers for IDR: the nature of societal problems and the growing complexity of research problems. They found that IDR is enabled by a number of dynamic characteristics of the scientific enterprise, including

- Disciplinary strength.
- Increased accessibility of data.
- Increased computing power.
- Increased power and accessibility of scientific instrumentation.
- Increased communication and the Internet.
- Ease of collaborating across institutional and programmatic borders.

The participants in the groups suggested a number of interesting models that sponsoring agencies could use to support IDR:

- Providing a mechanism to acknowledge collaborating investigators.
- Facilitating collaboration and agreements between and among institutions, including the national laboratories.
- Examining the need for the purchase, technical operation, and upgrading of large, shared instrumentation independent of individual projects.
- Breaking down of funding stovepipes within and between agencies.
- Interagency harmonization of award terms and conditions for similar research programs.
- Encouraging "grand challenges" or roadmaps.

---

[a]Alignment of Funding Mechanisms with Scientific Opportunities, October 27, 2003 Workshop Summary from NSTC's Regional Forum on Research Business Models *http:// rbm.nih.gov/afmso.html*.

[b]Gabriel, C. Comments at Convocation on Facilitating Interdisciplinary Research, January 29, 2004, Washington, D.C., *http://www7.nationalacademies.org/interdisciplinary/Convocation_Agenda.html*; "Ten Research Business Models Objectives Cleared by NSTC Science Committee." *The Blue Sheet*, 2004. 47(011):3.

*"We study, we plan, we research. And yet, somehow,
money still remains more of an art than a science."*

judgment of those who have extensive experience in interdisciplinary work
(see Box 8-5).

## CONCLUSIONS

Funding organizations at all levels and of all sizes have great opportunities to facilitate both disciplinary and interdisciplinary research (see Box
6-9). Indeed, some of them have been pioneers in promoting steps suggested
in this report, such as creating special IDR initiatives that can be critical to
the evolution of a vital but complex field.[16]

Some funding organizations have also recognized that research fields
and methods are now so interdependent that it may not be possible to fund
"just microbiology" or "just physics." Instead, they have found it desirable,
in addressing objectives in some fields, to support a wide framework of
disciplines simultaneously. For example, to support a program in the life
sciences, an organization may have to fund mathematics, probability, chemistry, computer science, biomedical engineering, and other relevant fields,
as well as biology.

By extending and adapting procedures developed earlier to evaluate
research proposals for single-discipline topics, funding organizations may

---

[16]An example is the rapid effort by NIH to launch a program of vaccine development
against agents of bioterrorism.

## EVOLUTION

### BOX 6-9 The Emergence of Biomedical Engineering: A Case Study in Collaboration among Researchers, Societies, and Funders[a]

The roots of biomedical engineering[b] reach back over 200 years to early developments in electrophysiology. Biomedical engineering has evolved through the collaboration of engineers and clinical scientists. The profession has been characterized by the emergence of separate societies with a focus on field-specific applications. As a step toward unification, an umbrella organization, the American Institute for Medical and Biological Engineering,[c] was created in 1992.

The earliest academic programs began to take shape in the 1950s. In the early 1960s, the National Institutes of Health (NIH), petitioned by researchers to develop educational programs for bioengineers, took three steps to support the emerging field. It created a program-project committee under the National Institute of General Medical Sciences to evaluate program-project applications, many of which served biophysics and biomedical engineering. Then it set up a biomedical engineering training study section to evaluate training-grant applications, and it established two biophysics study sections. A special "floating" study section processed applications in bioacoustics and biomedical engineering.

The field received a large push when The Whitaker Foundation[d] was created in 1975. In 1992, the Whitaker Foundation initiated large grant programs designed to help institutions to establish or develop biomedical engineering departments or programs. By 2002, Whitaker had contributed more than $615 million to universities and medical schools to support faculty research, graduate students, program development, and construction of facilities.

The National Science Foundation (NSF) and NIH, individually and collaboratively, have helped to provide a structure for research efforts. NSF established the Biomedical Engineering Division in the Directorate of Engineering in 1990. In 1991, NIH and NSF set up a collaborative workshop on biomedical engineering training.[e] The NIH director established the Bioengineering Consortium[f] in 1997, and in 2000 the National Institute of Biomedical Imaging and Bioengineering (NIBIB)[g] was created by Congress.

---

[a]History of Biomedical Engineering. Whitaker Foundation Web site. *http://www.whitaker. org/glance/history.html.* Accessed April 30, 2004.

[b]Bioengineering integrates physical, chemical, mathematical, and computational sciences and engineering principles to study biology, medicine, behavior, and health. It advances fundamental concepts; creates knowledge from the molecular to the organ systems levels; and develops innovative biologics, materials, processes, implants, devices, and informatics approaches for the prevention, diagnosis, and treatment of disease, for patient rehabilitation, and for improving health. NIH Working Definition of Bioengineering. July 24, 1997. *http:// www.becon2.nih.gov/bioengineering_definition.htm.*

[c]AIMBE home page *http://www.aimbe.org/.* Accessed April 30, 2004.

[d]Whitaker Foundation home page. *http://www.whitaker.org/.* Accessed April 30, 2004.

[e]*Summary of the NIH/NSF Workshop on Bioengineering and Bioinformatics Research Training and Education* (June 13-14, 2001) *http://www.nibib.nih.gov/training/NIHNSF/NIHNSF Training.pdf.*

[f]BECON home page *http://www.becon2.nih.gov/becon2.htm.* Accessed April 30, 2004.

[g]NBIB home page *http://www.nibib.nih.gov/.* Accessed April 30, 2004.

be able to overcome some important current barriers to IDR. Funding organizations can be most effective when they engage in extensive dialogue with leading practitioners to learn where the opportunities are greatest.

## FINDING

The characteristics of IDR pose special challenges for funding organizations that wish to support it. IDR is typically collaborative and involves people of disparate backgrounds. Thus, it may take extra time for building consensus and for learning of new methods, languages, and cultures.

## RECOMMENDATIONS

### Funding Organizations

**F-1: Funding organizations should recognize and take into consideration in their programs and processes the unique challenges faced by IDR with respect to risk, organizational mode, and time.**

For example, funding organizations can seek to

- Ensure that a request for proposals does not inadvertently favor funding a single-discipline project over an IDR project; for example, by including limitations on funding amounts, duration of funding (successful IDR teams often take longer to build and to coalesce), scope, and allowable travel and other budget items, all of which would militate against IDR.
- Develop funding programs specifically designed for IDR, for example, by focusing research around problems rather than disciplines.
- Provide seed-funding opportunities for proof-of-concept work that allows researchers in different disciplines to develop joint research plans and to perform initial data collection or for new organizational models or project approaches that enable IDR.
- Provide support for universities for shared research buildings, large equipment, or specialized personnel (machinists, glassblowers, and computer and electronic technicians).
- Provide funding mechanisms that allow researchers to obtain training in new fields.
- Fund programs of sufficient duration to allow for team-building and integration of research efforts.
- Provide funding mechanisms so that universities (including those from different countries) can work together to address societal problems that each would be challenged to address alone.

Develop mechanisms for budgetary flexibility in long-term, multi-institutional grants.

- Acknowledge, for projects that require more than a single principal investigator (PI), the equal leadership status of multiple PIs when "co-PI" is ambiguous.
- Remove administrative barriers to, and explicitly encourage, partnerships between universities, industry, and federal laboratories to facilitate IDR.

**F-2: Funding organizations, including interagency cooperative activities, should provide mechanisms that link interdisciplinary research and education and should provide opportunities for broadening training for researchers and faculty members.**

They can

- Require institutions that receive IDR funding to demonstrate support for interdisciplinary educational activities, such as team teaching.
- Provide, to the extent allowed by the funding organization's mission and guidelines, special grants to support interdisciplinary teaching.
- Designate funds for IDR meetings that encourage interaction between researchers in different disciplines so they can learn about the research in other fields and network with other researchers with whom they might collaborate.
- Support sabbaticals and leaves of absence for studies that focus on interdisciplinary scholarship.
- Ensure that their staff is knowledgeable about interdisciplinarity.

**F-3: Funding organizations should regularly evaluate, and if necessary redesign, their proposal and review criteria to make them appropriate for interdisciplinary activities.**

For example, funding organizations can

- Develop criteria to ensure that proposals are truly interdisciplinary and not merely adding disciplinary participants.
- Encourage IDR proposals that fall within the compass of the organizations' overall missions even if they cross internal organizational boundaries or do not fit specific (review) divisions.
- If they are organized along disciplinary lines, develop policies and practices for funding research that may have a major impact on research in other disciplines, for example, by awarding a mathematics section grant to a mathematician to work on a life-sciences project.

F-4: Congress should continue to encourage federal research agencies to be sensitive to maintaining a proper balance between the goal of stimulating interdisciplinary research and the need to maintain robust disciplinary research.

# 7

# The Role of Professional Societies

M any professional, or disciplinary, societies were founded to support the single disciplines for which they are named.[1] And yet in recent decades, these societies, like many other organizations, have been increasingly called on to expand their relationships to new fields of research. In addition, a new breed of professional society has arisen, mostly after World War II, that is primarily interdisciplinary (see Appendix D). Among the many interdisciplinary societies are the IEEE Computer Society (1946), the Society of Industrial and Applied Mathematics (1952), the Biophysics Society (1956), the Biomedical Engineering Society (1968), and the Materials Research Society (1973) (see Figure 7-1).

The mission of the professional societies is primarily educational and informational. Their influence flows from their continuing and highly visible functions: to publish professional journals, to develop professional excellence, to raise public awareness, and to make awards. Through their work, they help to define and set standards for their professional fields and to promote high standards of quality through awards and other forms of recognition.

---

[1]Among the oldest professional societies are the American Society of Civil Engineers (1852), American Chemical Society (1876), American Mathematical Society (1888), and American Physical Society (1899).

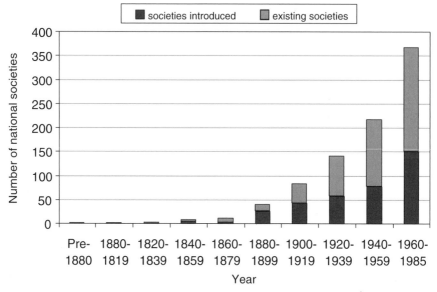

**FIGURE 7-1** Growth in numbers of professional societies, 1880-1985.
NOTES: Many national professional associations were founded over the period 1880-1985; founding dates are grouped into 20-year periods.
SOURCE: The data are from the *Encyclopedia of Associations, 1985* as compiled by Burton R. Clark in *The Academic Life: small worlds, different worlds* (1987).

## A VISION FOR PROFESSIONAL SOCIETIES
## THAT WISH TO FACILITATE IDR

In some ways, professional societies have a clearer overview of trends in their fields than do federal agencies, universities, and funding organizations. The central position of professional societies brings excellent leverage with which to design and promote change, including through publications, policy statements, meetings, committees, lectureships, and awards.

One particularly important function of professional societies relative to IDR—publishing professional journals—is shared with commercial publishers, some of which are large and influential forces in their own right. Because commercial publishers are for-profit ventures, however, their mission differs in an important way from that of the societies, which address the full gamut of concerns and achievements of their scientist and engineer members.

## PUBLICATION BARRIERS ENCOUNTERED BY RESEARCHERS

With the exception of a few leading general journals—such as *Science*, *Nature*, and the *Proceedings of the National Academy of Sciences*—the prestigious outlets for research scholars tend to be the high-impact, single-discipline journals published by professional societies. Although the number of interdisciplinary journals is increasing, few have prestige and impact equivalent to those of single-discipline journals, so students and faculty who publish in them might not receive the recognition they need for professional advancement.

Interdisciplinary researchers may find some recognition by publishing in single-discipline journals (journals to which part of their work is relevant), but the truly integrated portion of their research may not be clear too much of the audience or be noticed by peers who do not read those journals.

A general concern of researchers is the need to produce evidence of appropriate productivity at the time of tenure review. Members of a review panel usually want to know which journals a researcher publishes in and what impact those publications have had on other researchers. A person working on an interdisciplinary project may be publishing in an interdisciplinary journal that is unfamiliar to some reviewers.

## SUPPORT FOR PEOPLE AND PROGRAMS

The policies of professional publications have a strong influence on researchers who must gain public recognition to advance in their careers. Those published by professional societies have an opportunity to lower barriers to researchers by revising policies. In the committee's survey, the top two recommendations for journal editors were to incorporate interdisciplinary expertise in review panels (38.8 percent) and to feature novel innovations and initiatives (36.2 percent); 17.3 percent of respondents reported that they were satisfied with the current situation.

### Research Publications

Disciplinary societies have a great deal of influence through their journals in terms of their willingness to publish, their review procedures for papers submitted to a journal, and their ability to create new journals for subdisciplines. In addition, disciplinary society newsletters can be used to facilitate communication among disciplines (see Box 7-1).

Disciplinary societies could help their members by founding or promoting new journals, new sections, and other kinds of homes for emerging interdisciplinary subjects. They can also help researchers by giving awards

## TOOLKIT

### BOX 7-1   The Role of Journals in Fostering IDR

**Publishing journals that support interdisciplinary research**

The creation of journals that are dedicated to publishing research at the intersection of two or more fields is critical to the development of IDR and is another way that societies can foster this type of investigative approach.

Here is a selected list of recently established journals that represent research fields arising from the combination of disciplines:

- Archaeoastronomy (University of Texas Press for The Center for Archaeoastronomy in cooperation with ISAAC, the International Society for Archaeoastronomy and Astronomy in Culture)
  - Astrobiology (Mary An Liebert)
  - Biogeochemistry (Kluweronline)
  - Computation Geosciences (Kluweronline)
  - Ethnomusicology (Society of Ethnomusicology)
  - Internet Mathematics (AK Peters)
  - Journal of Neuroscience (Society for Neuroscience)
  - Neuropsychopharmacology (American College of Neuropsychopharmacology)
  - Geochemistry, Geophysics, Geosystems (American Geophysical Union)
  - Transactions in Mechatronics (IEEE/ASME)

The emerging field of bioeconomics has the *Journal of Bioeconomics,* which was created in 1999 "to encourage alternative approaches and creative dialogues between economists and biologists and a transfer of concepts, theories, and tools, and data bases in both directions by extending and integrating economics and biology." The journal "is interdisciplinary in spirit and open to various schools of thought and methodologies."[a]

**Highlighting important research in other fields**

In some journals, there is a regular and committed effort to expose readers to research and news in other fields. For example, the journal Cell includes a section that highlights recent findings on signaling mechanisms with article summaries from such journals as *Neuron, Immunity, Molecular Cell,* and *Current Biology* and another section that highlights cancer biology findings with article summaries from such journals as *Cancer Cell, Current Biology, Immunity,* and *Chemistry & Biology.* The interdisciplinary journal *Science* has a regular feature called "Editor's Choice" that highlights recent publications in many fields and journals.

---

[a]Landa, J. and Ghiselin, M. 1999. "The Emerging Discipline of Bioeconomics: Aims and Scope of the Journal of Bioeconomics." *Journal of Bioeconomics* 1(1):5-12.

and recognition for interdisciplinary work; this would help faculty who are working on interdisciplinary projects and who must demonstrate the value of their work to review committees that might not be familiar with either the interdisciplinary field or the interdisciplinary journals of significance to it.

Mathematics journals, for example, traditionally discourage researchers from submitting papers on interdisciplinary research; this tends to be true as well in chemistry. But some journal editors have broken with tradition to publish papers that turn out to have high importance to emerging fields. An example is an early paper on string theory by Edward Witten, a theoretical physicist, in 1983. The paper was published in *Communications in Mathematical Physics* (Arthur Jaffe, editor)—primarily a mathematics journal—over considerable objection. A decade later, Witten was awarded the Fields Medal in mathematics, and the interdisciplinary pursuit of string theory is today of major importance for both mathematics and physics.

In addition, some societies create a subscription model based on article access rather than journal title. For example, IEEE allows access to all its professional journals, regardless of which subdiscipline's journal or conference published an individual article.

## Program Initiation

Societies have taken and will continue to take a direct and active role in initiating IDR programs (see Box 7-2). In particular, the sponsorship of interdisciplinary groups (such as biochemistry in a chemistry society or biogeochemistry in a geophysical society) may constitute a proving ground for new disciplines as they emerge.

In addition, societies are able to

- Award prizes to students and faculty for excellent IDR proposals or projects. Such awards and other professional recognition can be important in helping an interdisciplinary researcher to gain tenure. For example, the American College of Neuropsychopharmacology gives many such awards that can enhance the careers of researchers in a field of considerable complexity. In the American Geophysical Union, all disciplinary awards include a sentence indicating that IDR investigators can qualify for the award.
- Target stipends, awards, or scholarships that permit students to spend time in other laboratories or with collaborators at various institutions.
- Invite interdisciplinary experts to serve on standing committees when that is appropriate.
- Reward outstanding mentors of interdisciplinary activities.

---

**TOOLKIT**

## BOX 7-2  Professional Societies Have Fostered IDR through a Number of Initiatives

### Hosting Workshops on Emerging Subjects

Professional societies often host seminars, meetings and colloquiums that bring together scientists in different disciplines to learn about diverse fields and research topics, to learn the languages of different fields, and to discover where these research topics overlap. In December 2000, for example, the American Academy of Microbiology held a colloquium titled "Geobiology: Exploring the Interface Between the Biosphere and the Geosphere."[a] The colloquium participants outlined a number of challenges facing this emerging field and called for interdisciplinary training of researchers and funding of research projects in the new field. Similarly, both the American Geophysical Union and the Geological Society of America have held special sessions on geobiology at their annual meetings since 2000.

### Organizing Interdisciplinary Society Panels or Divisions

Organizations that support researchers often foster IDR through the creation of groups or divisions in the society. For example,

• In 2000, the American Geophysical Union formed a section on Biogeosciences.[b] The Geological Society of America created the Geobiology and Geomicrobiology Division in May 2001.
• The Institute of Electrical and Electronics Engineers has an Engineering in Medicine and Biology Society, which has a Sensors Council.
• The American Institute of Chemical Engineers has a Food, Pharmaceutical, and Bioengineering Division.

There is also an increasing trend toward intersociety collaborations, such as the joint meetings on interdisciplinary topics sponsored by SIAM (Society for Industrial and Applied Mathematics) and the American Statistical Association.

### IDR Awards

A disciplinary society can support IDR by granting research awards. The Fund

---

All such steps can focus more attention on IDR and strengthen the reputation of IDR among academic institutions and funding organizations.

## SUPPORT FOR IDEAS AND INITIATIVES

After publications, the second important forum of professional societies is their regional and national meetings. By bringing the right people

for the Advancement of the Discipline[c] awarded by the American Sociological Association "provide scholars with venture capital that has the potential for challenging the discipline, stimulating new lines of research, and creating new networks of scientific collaboration." It provides up to $7,000 in unrestricted funds "to provide opportunities for substantive and methodological breakthroughs, broaden the dissemination of scientific knowledge, and provide leverage for acquisition of additional research funds." Recent winners include Charles Kurzman of the University of North Carolina-Chapel Hill for a series of workshops to bring together scholars in three overlapping fields (Islamic movements, social movement studies, and social network analysis) in two sets of workshops designed to stimulate intellectual cross-fertilization among them; and Marjorie L. DeVault of Syracuse University for a conference that will bring together distinguished senior scholars, mid-level scholars, and graduate students to develop Institutional Ethnographic (IE) approaches for studying the workings of economic restructuring.

### Interdisciplinary Recognition Awards and Lectureships

Professional societies can recognize and encourage IDR by granting awards to researchers whose interdisciplinary work has advanced the field. For example, at its annual meeting, the Materials Research Society presents one member with the von Hippel Award for "brilliance and originality of intellect, combined with vision that transcends the boundaries of conventional scientific disciplines."[d] Esteemed lectureships, such as the George A. Miller lectureship of the Cognitive Neuroscience Society, can also highlight researchers' IDR and provide a venue for recipients to describe their work to others outside the field.[e]

---

[a]Geobiology: Exploring the Interface between the Biosphere and the Geosphere, Colloquium Report 2001. American Society for Microbiology, available at *http://www.asm.org/Academy/index.asp?bid=2132.*

[b]AGU Adds Biogeosciences Section. AGU Press Release No. 00-16, June 8, 2000. *http://www.agu.org/sci_soc/prrl/prrl0016.html.*

[c]Fund for the Advancement of the Discipline, available at *http://asanet.org/members/fad.html.*

[d]The von Hippel Award of the Materials Research Society, available at *http://www.mrs.org/awards/VonHip.html.*

[e]Information available at the Cognitive Neuroscience Society Web site at *http://www.cogneurosociety.org/content/February%202003.*

together, these meetings and the activities that grow out of them can nourish new ideas and initiatives.

## Professional-Society Meetings

Meeting organizers have opportunities to devise many kinds of strategies that promote interdisciplinary research and education. Society meetings are effective venues for interdisciplinary researchers to get together with potential collaborators, interested employers, and sympathetic institutions (see Box 7-3). Searches for interdisciplinary positions can be facilitated through formal presentations, informal drop-in rooms, and coffee sessions. Funding agency representatives can discuss grant mechanisms and topics of high funding priority, allowing graduate students, postdoctoral scholars, and faculty to plan programs and partnerships. Organizers can hold topical interdisciplinary symposiums or colloquiums that are sponsored jointly by other societies.

### Promoting the Integration of Disciplines

Societies can plan special activities to facilitate communication between disciplines. They can form alliances with other professional societies to help researchers in different disciplines to become more familiar with one another and one another's research (see Box 7-4). To help to encourage familiarity, they can develop a lexicon that explains the vocabulary of the field in general scientific terms. Communication becomes more important as some older disciplines become more interdisciplinary; for example, biogeosciences recently became the subject of a new section of the American Geophysical Union. In addition, they might offer joint awards with other associations.

*"So NOW I understand what that means."*

---

INNOVATIVE PRACTICE

**BOX 7-3   The Association of American Geographers**

The Association of American Geographers (AAG)[a] celebrated its 100th anniversary in 2004. The centennial meeting was attended by over 5,000 people and showcased many of the ways that the society has supported integrative research and partnerships with other disciplines and organizations. More than 3,000 papers and posters were presented on a wide variety of geographic topics, many of them interdisciplinary and many by scholars from outside the discipline.

Plenary sessions featured internationally renowned scholars in a variety of disciplines who spoke of their own research and of their perception of geography and its role in interdisciplinary education and research. Speakers included National Academies President Bruce Alberts, a biochemist; past National Science Foundation Director Rita Colwell, a microbiologist; Clark University Professor Cynthia Enloe, a political scientist; and Columbia University Professor Jeffrey Sachs, an economist. In addition, many of AAG's 54 specialty groups invited speakers in related disciplines with financial support from the AAG Enrichment Fund, especially established for the purpose.

AAG presents an annual honorary geography award to recognize contributions to geographic knowledge by scholars outside the discipline. In 2004, the award went to Georgetown University Professor of Decision Sciences Keith Ord, a pioneer in spatial statistics. The centennial meeting included workshops that highlighted core interdisciplinary research tools, such as geographic information systems, global positioning systems, and other new technologies for integrating and analyzing spatial data from multiple disciplines. Other sessions focused on collaborative public, private, and academic partnerships for research, education, and outreach, including the My Community, Our Earth (MyCOE) project to help middle- and high-school students around the world learn to study and propose solutions to sustainable-development issues in their own communities.

---

[a]Association of American Geographers Web page *http://www.aag.org/*.

---

Other opportunities in communication and education include initiatives to

- Provide journal subscriptions at reduced cost to members of other societies.
- Cosponsor sessions at the main meetings of other societies.
- Jointly sponsor workshops, other small topical meetings, and field trips.
- Offer short courses at other meetings.
- Cooperate with other societies on K-12 and undergraduate educational programs.

---

**INNOVATIVE PRACTICE**

**BOX 7-4   Models for Collaboration between Professional Societies**

Often, societies that support individual disciplines recognize the importance of interdisciplinary collaboration. Because of the representation of diverse populations of researchers in their membership, societies hold a unique position: they can be the collective mouthpiece of researchers and identify research subjects or fields that are weaker both in numbers of people involved and in the current body of knowledge. When societies team up to try to strengthen these neglected subjects or fields, they build on the experience of their society members and hold a dialogue that includes people in different fields (see Box 6-9).

Recently, the Coalition for Bridging the Sciences[a] identified the interface of biology with physics, mathematics, engineering, and computer science as having great potential. The coalition is made up of nine research societies that represent 126,000 scientists in academe and industry. Members of each society emphasized the importance of their disciplines in the progress of biomedical research. In this specific example, the member societies called for a review of federal funding of these "supporting" disciplines and asked that a new funding entity be created to focus on long-term research in subjects not covered by existing funding mechanisms.[b] In particular, it would support basic research to develop technology and innovations necessary for the advance of biomedicine.

The synergy of research societies in fostering IDR is powerful. Disciplinary societies are a convenient medium for researchers to voice their opinions, and such initiatives as regular focus groups on emerging research concerns can promote the recognition of topics ripe for interdisciplinary collaboration. The collaboration of disciplinary societies can allow their members to interact and develop a common language and to learn more about research in other fields. Finally, collaboration between societies can have more influence on the support of IDR topics because they present a unified front that comprises the memberships of the participating societies.

---

[a]http://www.biophysics.org/pubaffair/bsc.htm.
[b]Couzin, J. Congress wants the Twain to Meet. *Science* 301:444, 2003.

---

- Publish special issues of periodicals independently or jointly with other societies.

## SUPPORT FOR INSTITUTIONS AND FACILITIES

Although the missions of professional societies do not ordinarily include direct support for institutions in which research is performed, the societies strongly influence practices and attitudes related to IDR.

## Developing Norms for Interdisciplinary Activities

One of the overarching needs for research is better mechanisms to evaluate the quality and success of interdisciplinary activities. Professional societies can be leaders in proposing and developing norms for interdisciplinary practice. For example, they might suggest appropriate skills and standards that should be mastered by students and faculty who participate in interdisciplinary research and education. They could publicize practices found to promote success, such as inclusion of funding in research grants to support substantial startup time during which participants can absorb the language and culture of multiple disciplines. They can invite the members of successful IDR teams to write or talk about their experiences in the society journals and at meetings.

## CONCLUSIONS

Most researchers are members of professional societies. When these societies choose to support a particular policy, they convey the "voice" of the research enterprise with a unique degree of legitimacy. They now have the opportunity to raise that voice on behalf of interdisciplinary research and education: to broaden the interdisciplinary outlook of scientists, to recognize young interdisciplinary scientists of talent, and to facilitate the interdisciplinary strengths of their society.

## FINDING

**Professional societies have the opportunity to facilitate IDR by producing state-of-the-art reports on recent research developments and on curriculum, assessment, and accreditation methods; enhancing personal interactions; building partnerships among societies; publishing interdisciplinary journals and special editions of disciplinary journals; and promoting mutual understanding of disciplinary methods, languages, and cultures.**

## RECOMMENDATIONS

### Professional Societies

**PS-1: Professional societies should seek opportunities to facilitate IDR at regular society meetings and through their publications and special initiatives.**

For example, societies can

- Include IDR presentations and sessions at regular society meetings by

— Choosing IDR topics for some of the seminars, workshops, and symposia.

— Promoting networking and other opportunities to identify potential partners for interdisciplinary collaboration.

— Cohosting symposia with other societies.

— Holding workshops on communication skills, leadership, consensus-building, and other skills useful in leading and being part of IDR teams.

- Establish special awards that recognize interdisciplinary researchers.

- Sponsor lectureships that bring recognition of the value of interdisciplinary experience.

- Prepare glossaries, primers, tutorials, and other materials to assist scientists in other fields who wish to learn new disciplines.

- Create sections, divisions, or boards that represent interdisciplinary aspects of their fields.

### Journal Editors

**J-1: Journal editors should actively encourage the publication of IDR research results through various mechanisms, such as editorial-board membership and establishment of special IDR issues or sections.**

In particular, journal editors can

- Increase the exposure of IDR by devoting special issues or sections to specific IDR directions in a field and accepting more research papers that introduce new IDR areas.

- Add researchers with interdisciplinary experience to editorial boards and review panels and develop specific techniques for evaluating interdisciplinary submissions.

- Consider whether their publications' guidelines for authorship and submission of manuscripts are appropriate for IDR.

- Take steps to improve the sharing of knowledge between disciplines by publishing

— Comprehensive review articles on related disciplines.

— Overview articles on fields relevant to published interdisciplinary works.

— A list of the fields covered in interdisciplinary papers.

— Hyperlinked text in papers directing on-line readers to discipline-specific educational resources.

- Create subscription models based on article title and subject rather than journal title to enhance cross-discipline access.

# 8

# Evaluating Outcomes of Interdisciplinary Research and Teaching

F unding organizations and academic institutions need effective ways to evaluate the outcomes and effects of their investments in interdisciplinary research and teaching, just as they do for disciplines, to determine whether their goals are being achieved.

Appropriate evaluation is critical not only to assess the outputs of research but also to view more general outcomes in terms of organizational goals. For example, in evaluating a federal program to reduce unemployment, policy makers are less concerned about the *output*: how many people participate in the program, and more so by the *outcome*: how many participants obtain employment as a consequence of participating in the program within a particular period. The same is true of research: there is less interest in the number of publications than in the impact of the publications in terms of their quality, relevance, and stature.

## THE CHALLENGE OF EVALUATING RESEARCH

As discussed in the Committee on Science, Engineering, and Public Policy (COSEPUP) report *Evaluating Federal Research Programs*,[1] the useful outcome of interdisciplinary or disciplinary basic research cannot be measured directly on an annual basis because of its inherent unpredictability.

---

[1]National Research Council. 1999. *Evaluating Federal Research Programs: Research and the Government Performance and Results Act.* Washington, D.C.: National Academy Press.

*149*

But that does not mean that no meaningful measures exist. COSEPUP found that measures do exist: measures of *quality*, in terms of research advancement; *relevance*, in terms of application development; and *leadership*, in terms of the ability to take advantage of opportunities when they arise, as evaluated by experts and users of research. In addition, COSEPUP concluded that human-resource development is also a key outcome of an effective research program.

A remaining challenge is to determine what additional measures, if any, are needed to evaluate interdisciplinary research and teaching beyond those shown to be effective for disciplinary activities. Successful outcomes of an interdisciplinary research (IDR) program differ in several ways from those of a disciplinary program. First, a successful IDR program will have an impact on multiple fields or disciplines and produce results that feed back into and enhance disciplinary research. It will also create researchers and students with an expanded research vocabulary and abilities in more than one discipline and with an enhanced understanding of the interconnectedness inherent in complex problems.

"I ADMIRE THE INQUIRING MIND AND THE PRAGMATIC MIND, BUT I ALSO ADMIRE SOMEONE WHO CAN HIT."

The remainder of this chapter examines the challenges of evaluating interdisciplinary research and teaching and provides examples of innovative techniques (see Box 8-1). The report does not presume to prescribe specific evaluation measures; that is best done by each institution involved in IDR on the basis of its own objectives and culture. The examples cited here are intended to demonstrate approaches that may be useful.

---

**TOOLKIT**

**BOX 8-1 Measures to Evaluate Interdisciplinary Work**

In a recent study by the Harvard Interdisciplinary Studies Project, Veronica Boix Mansilla and Howard Gardner looked at research and teaching practices at several interdisciplinary institutes and programs over the last 2 years. They focused on the appropriate ways to evaluate IDR by interviewing over 60 researchers. They found that researchers typically were judged on indirect or field-based measures of quality, such as numbers of patents, publications, and citations; the prestige of universities, funding agencies, and journals; and approval of their peers.

Interdisciplinary research varies broadly in specific goals and validation criteria. Researchers at the cutting edge find that they also have to develop criteria with which to gauge their progress. On the basis of these interviews, Mansilla and Gardner suggest that measures of acceptability directly addressing the substance of interdisciplinary work be considered together:

**1. The degree to which new interdisciplinary work relates to antecedent disciplinary knowledge.** Even though engaged in interdisciplinary work, researchers still evaluated the credibility of new findings on the basis of consistency with the "disciplinary canon"—often in more than one field. High-quality understanding required more than a sum of disciplinary rules—it required a "unique coordination of disciplinary insights."

**2. The sensible balance reached in weaving perspectives together.** The interviewees appreciated interdisciplinary work that thoughtfully balanced perspectives of the disciplines represented, even though disciplinary standards could conflict with regard to worthwhile topics of inquiry or measures of proof.

**3. The effectiveness with which a particular piece of work advances understanding and inquiry.** Among interviewees, contributions oriented toward pragmatic problem solving and product development placed a premium on standards of viability. Algorithmic models of complex phenomena were associated with measures of simplicity and predictive power. Multidimensional phenomena were evaluated on the basis of comprehensiveness and empirical grounding.

---

[a]Mansilla, V.B. and Gardner, H. "Assessing Interdisciplinary Work at the Frontier: An empirical explanation of symptoms of quality." *http://www.interdisciplines.org/interdisciplinary/ papers/6/2/printable/paper.*

## EVALUATING RELATIVE TO THE DRIVERS OF IDR

Chapter 2 discusses the four driving forces of IDR:

- The inherent complexity of nature and society
- The drive to explore the interfaces of disciplines
- The need to solve societal problems
- The stimulus of generative technologies

One way to evaluate IDR is to consider it in light of those forces. For example, does a given program deal with the inherent complexity of nature and society that must be addressed by multiple disciplines? How well does the program do that? Each funding organization, depending on its own mission goals, would expect to use a different combination of drivers for its evaluation.

The drive to explore the *interfaces of disciplines* could be evaluated by examining the extent to which researchers truly collaborate with other researchers in adjacent or complementary fields or stimulate the development of a new field. Especially relevant to earlier COSEPUP reports is the driving force of the need to *solve societal problems*, which usually involve at least some applied research. One measurable outcome of research generated by societal problems would be a practical answer to the original question. For example, an IDR effort to reduce hunger could measure practical progress toward that goal. The same program, of course, might produce additional outcomes of value, including basic research, that were not anticipated.

The *stimulus of generative technologies* could be evaluated by examining the degree to which new technologies are developed that enhance research capabilities in many fields through the development of new instrumentation or informational analysis.

## EVALUATING THE DIRECT AND INDIRECT IMPACTS OF IDR

Many of the standard means for evaluating disciplinary research and teaching can also be applied to interdisciplinary research and teaching: the use of metrics, such as number of publications, citations of publications, and successful research-grant proposals; teaching evaluations by students; benchmarking with other programs (when comparable programs exist); and national or international awards for and recognition of researchers or teachers. However, IDR can be expected to have measurable outcomes in multiple elements of technique, theory, and application. Taking account of that expectation will require new evaluation criteria that match the cross-cutting nature of IDR. Both direct and indirect outcomes could be amenable to evaluation.

Evaluating IDR productivity can also be complicated because, although in some situations IDR may take more time than disciplinary research, it may have a high degree of depth and importance of achievement. The contribution achieved by a research team may be more than the sum of the individual accomplishments.

### Direct Contributions of IDR to Knowledge

One way to evaluate IDR programs is to look for direct contributions in the form of new knowledge (see Box 8-2). Some IDR programs are so large that they stimulate new understanding in multiple fields. Examples are the Human Genome Project, the Manhattan Project, the broad effort to prove the theory of plate tectonics, global-climate modeling, and the development of fiber optic cable. A current example is the study of extremophiles—microorganisms that thrive under extreme chemical and physical conditions—as part of the emerging field of geomicrobiology. Their existence has influenced both biology and geology by expanding our notions about the origin of life on Earth (including the possibility of an extraterrestrial origin of life related to meteoric bombardment) and the limits of life on Earth (studied at deep-sea hydrothermal vents sustained by chemical synthesis). The existence of extremophiles has also altered traditional geochemical ideas about the formation and mediation of processes that lead to deposits of such ores as golds and sulfides.

Sometimes, the direct contribution of IDR is the creation of a new field or discipline as a result of the interactions between researchers who have a common interest. That was the case many years ago with biochemistry, and it is happening now in the formation of cognitive science, computational biology, nanoscience, and other fields (see Box 6-9 and Appendix D).

IDR may also add *value* to many traditional fields of research. For example, people studying nanoscience must bridge several disciplines seamlessly. Chemists are required for synthesis of nanostructures, materials scientists for characterization of structures, physicists for establishing new principles that relate quantum-like molecular states to new physical behavior on the nanoscale, and engineers for designing and building new devices and systems. At the same time, people use the richness of their nanoscience research experience to open up new *disciplinary* research directions and applications, such as the incorporation of nanostructures into bulk materials (see Box 6-6).

IDR may lead directly to the development of *new technologies* or *products*. Mathematical techniques developed for radiology now provide tools for oil companies to image the earth's upper crust. Researchers using prin-

## BOX 8-2  Evaluating IDR Center Proposals and Programs: The National Science Foundation Engineering Research Centers

The engineering research centers (ERCs) sponsored by the National Science Foundation (NSF) are systems-focused, interdisciplinary centers at universities all across the United States, each in close partnership with industry.[a] Primary goals of ERCs are to integrate engineering education and research, build competence in engineering practice, and produce engineering graduates with the depth and breadth of education needed for success in technological innovation and leadership of interdisciplinary teams.

NSF views ERCs as change agents for academic engineering programs and the engineering community at large. The mechanism of centers was chosen because centers can bring disciplines together. Since the ERC program was founded in 1985,[b] the ERCs collectively have brought substantial changes in the culture of academic engineering research and education.[c]

**Proposal Review**

ERC proposals are generated by program announcements.[d] Proposals are reviewed by a panel of peers selected by the NSF program manager. Panel members have scientific and technical expertise and experience in cross-disciplinary research, engineering education, industrial R&D, technology transfer, and research management. Panels recommend whether a prospective ERC should submit a full proposal. A second panel narrows the field and determines which sites to visit. The site-visit team consists of evaluators and two or three members of the panel. In 2002, there were 77 pre-proposals, 16 proposals, seven site visits, and four awards.

**Program Review**

An ERC begins operation with a 5-year award under a cooperative agreement with NSF. The agreement has the potential to extend NSF support to 10 years. After that, the ERC is expected to be self-sustaining. The progress and plans of each ERC are assessed annually through merit review by outside experts; review in the third year of operation can lead to extension of the cooperative agreement for 3 years to year 8, and a second review can take place in year 6. A period of phased-down support is provided to an ERC that is not renewed.

---

[a]Engineering Research Centers Association home page *http://www.erc-assoc.org/*; NSF, Division of Engineering Education and Centers, Engineering Research Center Program *http://www.eng.nsf.gov/eec/funding/pgm_display.cfm?pub_id=9971&div=eec*.

[b]The New Engineering Research Centers: Purposes, Goals, and Expectations. 1986. Washington D.C.: The National Academy Press.

[c]Suh, Nam P. The ERCs: What Have We Learned. *Engineering Education.* October 1987, p. 15-17. Engineering Research Centers Best Practices Manual *http://www.erc-assoc.org/manual/bp_index.htm*.

[d]Engineering Research Centers (ERC) Program Solicitation NSF 04-570 *http://www.nsf.gov/pubs/2004/nsf04570/nsf04570.htm*.

A team of peers similar to that constituted for proposal review conducts the third-year renewal review. Before the visit, the members of the site-visit team review the center's renewal proposal, using the ERC program performance-review criteria[e] available to all program participants. The first generation of ERC awards were evaluated on how well research needs were met for the industrial partner, 70 percent reported that participation in an ERC favorably affected their competitive position. Industry partners reported that the most important benefit was working with students.[f]

Renewal reviews are divided into six categories: systems vision and value added, strategic research plan, research program, education and educational outreach, industrial-practitioner collaboration and technology transfer, and strategic resource and management plan. Criteria in each category change in the three review periods—years 1-3, 4-6, and 7-10. For example:

| Evaluation Category | Years 1-3 | | Years 4-6 | Years 7-10 |
|---|---|---|---|---|
| | High Quality | Low Quality | High Quality | High Quality |
| Research program | Thrust team is appropriately cross-disciplinary, projects are becoming interdependent within the thrust and contributing to other thrusts | Thrusts and projects are single-disciplinary and isolated from one another, project results have no role in other projects and thrusts | Thrust team is appropriately cross-disciplinary, projects are interdependent within the thrust and contributing to other thrusts | Thrust team is appropriately cross-disciplinary, projects are interdependent within the thrust and contributing to other thrusts |
| Education & educational outreach | Cross-disciplinary research culture is developing, students work in teams, ratio of graduate to undergraduate is at most 2:1 | Little inter-dependence between faculty and students, involving few students in teams, ratio of graduate to undergraduate students is over 2:1 | Cross-disciplinary research culture has been developed, students work in teams, ratio of graduate to undergraduate students is at most 2:1 | Cross-disciplinary, team culture for students flourishes and impacts other parts of the university |

*continues*

---

[e]Third-Year Renewal Review Process and Review Criteria for Engineering Research Centers in the ERC Class of 2000. Engineering Research Centers Program, Division of Engineering Education and Centers, National Science Foundation. August 2002.

[f]Lynn Preston, Deputy Division Director (Centers), Division of Engineering Education and Centers, National Science Foundation. Staff interview conducted November 17, 2003.

---

**BOX 8-2   Continued**

**Outcomes**

The 41 ERCs have made substantial contributions to US industry, are leaders in developing interdisciplinary cultures in academe, and produce a wide array of knowledge and technological advances. Innovations in research management, education, precollege outreach, and technology transfer are documented by NSF and the Engineering Research Centers Association.[9]

---

[9]National Science Foundation, Division of Engineering Education and Centers, Engineering Research Center Program Achievement Showcase *http://www.erc-assoc.org/showcase/index.htm.*

---

ciples of molecular biology, nanofluidics, materials science, and engineering have been able to address well-identified medical and clinical needs, and this has led to progress in developing artificial tissues and drug-delivery systems.

### Indirect Contributions of IDR to Knowledge

Some contributions of IDR are less direct but substantial, and some institutions have begun attempts to evaluate them. For example, developing the engineering technologies necessary to achieve space flight has led to advances in the computer control of engineering processes, which have resulted in improvements in the reliability of industrial products and processes.

### Information-Sharing Networks

Researchers who divide their time between traditional disciplinary departments and interdisciplinary programs or centers often form "networks of practice"[2] through which they share information that does not always appear in immediate or traditional forms, such as publications in academic journals (see Box 8-3). Such information-sharing networks may yield other important outputs, such as congressional testimony, public-policy initiatives, mass-media placements, alternative-journal publications, and long-term product development.[3]

---

[2]Brown, J. S. and Duguid, P. *The Social Life of Information*, Cambridge, MA: Harvard Business School Press, 2002.

[3]Rhoten, D. 2004. "Interdisciplinary Research: Trend or Transition." *Items and Issues 5*, no. (1-2):6-11.

---

### EVOLUTION

### BOX 8-3 Social Network Evaluation of IDR Centers

The Hybrid Vigor Institute[a] conducted a social network analysis[b] of six interdisciplinary research centers. The object was to model the structure, relations, and positions of the research network in each of the centers, assess the relationships of the researchers in a given center, and identify "hot spots" of interdisciplinary activity.

Data were collected in two phases. First, social networks were evaluated using field survey and bibliometric methods. The survey determined a person's professional background with regard to disciplinary and interdisciplinary exposure, relationship with every other person in the center, and the center's organizational practices and processes. Second, researchers visited sites to collect observational data and perform interviews. The data were compiled and analyzed with a social network analysis.[c] Social network analysis provides useful insights into how well researchers in an interdisciplinary center interact with one another, and it can determine critical personnel for fostering collaboration. However, it does not match performance results with interactions.

Hybrid Vigor found that center networks were shaped by the diversity of and functional distance between the disciplines. There was a greater rate of connectivity among researchers of different disciplines than like disciplines; this suggests that researchers do seek interdisciplinary connections in the centers. In fact, on the average, 84 percent of the current connections were formed after the researchers joined the interdisciplinary center. Regardless of group size, researchers did not tend to interact with more than 15 other researchers. Position in the network is affected by professional rank and status; center directors act as nodes. Graduate students and postdoctoral scholars were connected with more people outside their discipline than were senior faculty. That indicates that, although center directors may act as the organizing force, it is graduate students and postdoctoral scholars that weave the web.

---

[a]Hybrid Vigor Institute home page *http://www.hybridvigor.org/.*

[b]Rhoten, D. Final Report, National Science Foundation BCS-0129573: A Multi-Method Analysis of the Social and Technical Conditions for Interdisciplinary Collaboration. September 29, 2003. Available at: *http://www.hybridvigor.net/interdis/pubs/hv_pub_interdis-2003. 09.29.pdf.*

[c]For a discussion of social network analysis, see National Research Council. *Dynamic Social Network Modeling and Analysis: Workshop Summary and Papers.* 2002. Washington, D.C.: National Academy Press.

## Quality of Educational Experience

One indirect impact of interdisciplinary programs is enrichment of the quality of undergraduate and graduate education. Interdisciplinary education programs have increased enrollments of undergraduate majors in IDR fields and enhanced non-majors' understanding of science and engineering.

- One example is an increase in earth-science enrollments at Stanford University that followed a shift in curriculum. After a substantial decline in the number of geology majors beginning in 1984, an interdisciplinary earth-systems degree was initiated in 1991-1992, and it led to a substantial increase in degrees awarded by the School of Earth Sciences (see Figure 8-1).

- Many universities have noted the popularity of science and engineering academic programs that are integrated with social-science issues. Two such programs are the Global Change Program of the University of Michigan[4] and the Program in Human Biology at Stanford University.[5]

- Experience at the University of Colorado at Boulder, a member institution of the National Center for Atmospheric Research, has shown that interdisciplinary programs that feature problems of greater breadth, societal relevance, or public policy (such as global change) are attracting more of the general student population to science courses. That is evidenced by the higher percentage of undergraduates who are taking more than the typical single required science course.

### Enhancing an Institution's Reputation

Another indirect impact of interdisciplinary research efforts and curricula is enhancement of an institution's reputation by establishing programs of high quality in cutting-edge, niche fields. That, in turn, can strengthen an institution's ability to attract outstanding graduate students, faculty, and postdoctoral scholars, as happened at the Joint Institute for Neutron Sciences at the University of Tennessee and Oak Ridge National Laboratory,[6] the Keck Graduate Institute,[7] and the Kavli Institute for Theoretical Physics at the University of California, Santa Barbara.[8]

### Demonstrating the Value of Instrumentation

IDR may demonstrate the value of a major tool or instrumentation that has multiple applications. For example, synchrotron radiation, which provides an ultrabright photon source, has had a major impact on many fields

---

[4]School of Natural Resources and Environment, Global Change Program home page *http://www.snre.umich.edu/faculty-staff-directory/list.php?unit_id=35.*

[5]Program in Human Biology home page, *http://www.stanford.edu/dept/humbio/.*

[6]Magid, Lee. Comments at Convocation on Facilitating Interdisciplinary Research, Washington, D.C., January 29, 2004. *http://www7.nationalacademies.org/interdisciplinary/Convocation_Agenda.html.*

[7]Keck Institute home page *http://www.kgi.edu/index_flash.shtml.*

[8]Kavli Institute home page *http://www.itp.ucsb.edu/.*

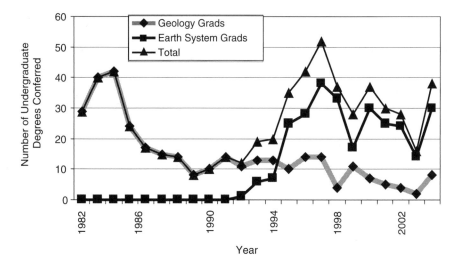

FIGURE 8-1    Degrees awarded by Stanford School of Earth Sciences.
NOTES: The undergraduate program at Stanford University, in an effort to offset a steady decline in student interest, created a new interdisciplinary earth sciences program. Student interest is reflected in the increased number of graduates declaring earth science as a major.

of research and on the development of industrial applications. It has also brought together multiple disciplines and groups of researchers, even though it was originally developed for the study of electronic and structural properties of materials. For example, in molecular biology, researchers use a synchrotron to obtain structures of proteins rapidly, and this enables pharmaceutical companies to develop new drugs (see Box 2-5).

## EVALUATING THE PEOPLE WHO PERFORM IDR

Many organizations would like to develop more effective ways to evaluate students and faculty who engage in IDR (see Box 8-4). One approach to such evaluation is to measure the degree to which steps suggested in Chapters 4 and 5 are implemented. When applied, those steps should have detectable effects on the success of students and faculty.

### Undergraduate and Graduate Students

Undergraduate and graduate students who work in more than one department might be expected to have experiences that they might not otherwise have; these experiences can provide starting points for evalua-

**TOOLKIT**

**BOX 8-4   Evaluating the NSF Integrative Graduate Education and Research Trainee (IGERT) Program**

The National Science Foundation (NSF) created the Integrative Graduate Education and Research Trainee (IGERT) program in 1997 in response to a growing recognition that graduate students in science and engineering needed to be better prepared for research that involves two or more disciplines. Since the first year of funding in 1998, NSF has added about 20 new programs each year. An institution that receives an IGERT grant currently receives up to $640,000 per year, the bulk of which is distributed as graduate student traineeships.

**Cross-Site Program Evaluation**

NSF has commissioned two cross-site reports to evaluate the impact of the IGERT program. The more recent focuses on the first two cohorts of IGERT projects in their third year of program implementation.[a] Data collection centered on interviews with students, faculty, and associated department chairs and university administrators. The key components evaluated were: project management, impact on students, impact on faculty, impact on institutions, and institutionalization.

**Self Assessment**

As described in the 2004 program solicitation, "IGERT projects are expected to incorporate and integrate . . . strategy and methodology for formative assessments of the project's effectiveness by individuals internal and external to the institution and program improvements based on these assessments." NSF used an

tion. The following examples may convey a sense of useful questions to ask of students in interdisciplinary programs:

- Are they working with and learning from students in other disciplines?
- Are they developing mastery of more than one discipline?
- Are they developing a sense of what it means to integrate more than one discipline in addressing a complex research question?
- Are they learning to use instrumentation or techniques that their own discipline might not provide?

### Postdoctoral Scholars

In a similar spirit, one might ask many of the same questions of postdoctoral scholars who are involved in IDR:

external evaluator to appraise assessment methods of 79 IGERT programs that were initiated between 1998 and 2001.[b] A direct survey of principal investigators (PIs) found that 71 percent were responsible for program assessment. In about half the programs, a committee of IGERT participants performed an assessment, whereas less than one-third of the programs relied on an external person or group to evaluate the program.[c]

The methods of assessment reported are listed below in order of prevalence:

- Informal feedback to PI or committee responsible for evaluation (84 percent).
- Annual or more frequent surveys of trainees (68 percent).
- Annual or more frequent meetings of project faculty members serving as an assessment committee (58 percent).
- Annual or more frequent meetings of project participants to discuss project management and problems in program implementation or function (57 percent).
- Survey of faculty for concerns (33 percent).
- Interview of participants and observation of classes, seminars, and laboratories by external evaluator (30 percent).
- Continuing observation by external evaluator (14 percent).

---

[a]IGERT Annual Cross-Site Report: 1998 and 1999 Cohorts. Fall 2003. Prepared by Abt Associates, Inc for NSF.

[b]IGERT Implementation and Early Outcomes: 2002. June 2003. Prepared by Abt Associates, Inc. for NSF.

[c]Kusmierek, K. and Pionte, M. "Content, Consciousness, and Colleagues: Emerging Themes from a Program Evaluation of Graduate Student Progress Toward Multidisciplinary Science." 42nd Annual Association of Institutional Research Forum. June 2002.

- Are they applying their own expertise in ways that add new value to a project and to their own grasp of one or more fields?
- Are they able to interact with specialists in other disciplines?
- Are they able to learn the language, content, and culture of another discipline?

## Faculty

Faculty who work in more than one department or discipline would be expected to receive many of the benefits sought by students and postdoctoral scholars, such as extending the range of their understanding, working on exciting topics at the frontiers of their field, and learning new disciplinary languages and cultures.

Evaluating the work or contribution of faculty who are participating in

IDR is not simple, however. For example, how does an academic department evaluate the radiochemist who has carried out substantial portions of a positron-emission tomography study but will not be the study's senior author? Each institution or funding organization that supports IDR is encouraged to devise ways of answering such questions.

On a more general level, the following questions might help to frame an evaluation:

- Are faculty members doing interdisciplinary work of high quality and reporting on it in leading journals or conferences?
- Are they working on topics that they might not otherwise be able to address in their original discipline?
- Have they extended their expertise in new directions?
- Have they participated in establishing new subfields?
- Do they include students or faculty from other disciplines in their own research work?
- Are their students successfully merging disciplines?
- Do they take part in multidisciplinary advisory or review groups?
- Have they been evaluated at their own institution by a multidisciplinary review group?
- Have they achieved recognition, such as awards and lectureships, for IDR or from another professional society outside their own field?
- Have they been invited to present work in venues outside their discipline (an interdisciplinary mathematician, for example, invited to give a presentation to a biology department or at a biological professional society)?

It is reasonable to assume that a series of such questions will point to a framework for evaluating faculty who are engaged in IDR.

## EVALUATING PROGRAMS, INSTITUTES, AND CENTERS THAT ENGAGE IN IDR

Many universities—motivated by the desire to organize work efficiently and to attract funds, students, and necessary infrastructure—have set up formal centers, programs, and institutes for IDR. Such structures are customary in industry and government, but their effectiveness in academe has not been thoroughly studied.

Indeed, the difficulty of developing effective review criteria is illustrated by a recent evaluation of NSF programs in IDR. Despite NSF's long-standing leadership, the evaluation urged the agency to "establish supplementary review criteria that will help to assess the quality of interdisciplinary effort in those programs where both single and multiple discipline proposals compete for a common pool of funds."[9] The report also stated that "no effective mechanism is in place to track or set performance goals for interdisciplinary research that can be used for planning, budget, and management decision-making."[10]

---

**Convocation Quote**

We need to do a better job of measuring results. . . . There is not a lack of data. There is a profusion of invisible data that needs to be better collected and disseminated.

Julie Thompson Klein, professor of humanities,
Wayne State University

---

At the heart of any evaluation process must be not only stringent peer review but also site visits that include personal interviews and objective observations. For most IDR programs, both internal and external reviews are essential to combine familiarity with institutional processes and objectivity of independent observation (see Box 8-5).

External review groups should represent all appropriate sectors; for example, in evaluating university centers, review groups should include the "users" of research outputs, such as industry, government, and policy representatives. To address the complexity of IDR, reviews should include mechanisms with two key qualities: depth of expertise in the core disciplines and related disciplines, and experience in carrying out IDR.

Recommendation of future directions for interdisciplinary centers should also include a "sunset" option. Initiatives will not be equally productive or equally long-lived. Reviewers should consider how much relevant new knowledge and understanding an IDR effort is generating and whether it should be terminated or moved in a new direction if the field itself changes.

For example, the discovery in the late 1980s of the fascinating $C_{60}$ molecule, with its icosahedral symmetry, attracted many researchers to study it and its associated cage molecules, collectively called fullerenes. After 5-8 years, much of the basic fullerene research had been accomplished; research priorities moved elsewhere, and centers closed or evolved. A good example of the latter situation occurred at Rice University, where the fullerene center successfully changed to a nanotechnology center (see Box 6-6). Addressing the complex strategic questions involved in identifying new directions and finding sufficient support can be aided by advisory committees that have the interdisciplinary expertise mentioned above.

---

[9]NAPA, ibid., 2004, p. ix.
[10]Ibid, p. 96.

**TOOLKIT**

## BOX 8-5 Assessment of Disciplinary and Interdisciplinary Research in the Netherlands

Assessment of scientific research at Dutch universities was started in the early 1980s. External review committees, consisting mainly of non-Dutch members, carry out evaluations on the basis of an evaluation protocol. The recently renewed assessment strategy was based partly on a model developed by the European Federation for Quality Management.[a]

### Self-Assessment

Self-assessment reports are to be written every three years and reviewed externally every 6 years. The self-assessment reports should contain several elements, of which the main ones are:

- Characterization of the institute: mission, formal collaborations, and affiliations.
- Leadership: organizational structure, list of research programs, and program leaders.
- Research strategy: organizational context, plans for short and long term.
- Researchers: personnel policy—selection, training, career planning, and mobility.
- Resources and funding: financial situation, research contracts, future funding prospects.
- Processes to support research: teamwork, supervision of PhDs, quality assurance.
- Reputation: expressed in, for example, citation scores, prizes, and awards.
- Internal assessment: monitoring of research management.
- External appreciation: dissemination of research outcomes.

Finally, academic institutions have begun to implement interdisciplinary courses and minors, and even a few majors, but development of course and program evaluations is in its early stages. The beginning questions to ask in framing an evaluation mechanism might include the following:

- Is interdisciplinary teaching attracting more of the general student population to science courses?
- Are interdisciplinary courses and programs attracting a new or different mix of students to careers in science?
- Are interdisciplinary courses effective vehicles for instilling science literacy and awareness of the roles of science and technology in modern life?

- Research outcomes: publications in refereed journals, and so on; patents.
- Future perspectives.

**External Review**

In the external assessment, four aspects are to be considered: the quality of the scientific research, the productivity of the scientific output (such as refereed publications), the relevance of the research for academe and society, and the future perspective, feasibility, and vitality of the research. In addition, it has proved useful to assess the research according to a five-point qualitative scale with scores, which are given for all four aspects separately:

5. *Excellent:* research that is internationally at the forefront and has a high impact.
4. *Very good:* research that is internationally competitive and nationally at the forefront.
3. *Good:* research that is nationally competitive and internationally visible.
2. *Satisfactory:* research that is solid but not exciting; nationally visible.
1. *Unsatisfactory:* research that is not solid or exciting; not worth pursuing.

**Interdisciplinary Themes**

For large interdisciplinary themes, the contributions by the different disciplines can be rated separately, for example, in written reports from experts in specific disciplines. The reports are offered to the assessment committee for final assessment. In a typical field, such as the biomedical and health sciences, in which multiple disciplines contribute (for example, physics, chemistry, biology, informatics, clinical medicine, and epidemiology), all research themes were assessed by a multidisciplinary committee, and the contributions of the different disciplines were assessed beforehand in writing by experts in those disciplines.

---

*a http://www.pgmm.org/efqm.htm.*

- Are students demonstrating a grasp of the complex interconnectedness of real-world problems?

## COMPARATIVE EVALUATIONS AND RANKINGS OF RESEARCH INSTITUTIONS

Comparative evaluations of research institutions, such as the National Academies' assessment of doctoral programs (see Box 5-1) and similar activities that rank university departments,[11] should include the contributions of both interdisciplinary activities and single-discipline contributions.

---

[11]For example, *U.S. News and World Report* issues annual department and program rankings.

For example, organizations evaluating such institutions can experiment with new forms of "matrix evaluation" to capture the activities and accomplishments of interdisciplinary researchers. A matrix approach would consider IDR as an integral part of the disciplines in which the IDR is "embedded" and make visible the cross-departmental efforts of people who form interdisciplinary teams.

A matrix-based evaluation might include in its criteria the comentoring of doctoral students, the contributions of people to multiple departments, and some publication criteria. Among their publication criteria might be the nature of the journal audiences for whom the work is published; citation analysis that reveals a broad interdisciplinary interest in the work being cited; double counting of publications, by which credit for a given paper is awarded to all coauthors; multiple authorship and coauthorship patterns that would reveal the disciplinary backgrounds of coauthors; and other measures that are still being developed.

## CONCLUSIONS

Better methods to evaluate IDR are needed to help funding organizations to assess the results of their investments better, to help sustain America's preeminence in higher education and research, and to enhance the contribution of IDR to the general advancement of science and engineering.

There has been little systematic study of the people, institutions, or funding organizations taking part in interdisciplinary activities. A few studies have begun, including the study by the Transdisciplinary Tobacco Use Research Centers (see Box 8-6), and their results will begin to add much-needed information to this nascent skill.

Despite the complexities of evaluating activities that span multiple disciplines, long experience in peer review and other assessment methods suggest that useful assessment techniques can be developed for IDR. This chapter has attempted to outline some of the topics to be studied and questions to be asked in constructing frameworks for evaluation. Given the inherent difficulty and expense of most interdisciplinary activities and the need to balance investments in research, it is essential to measure and maintain its value to the research enterprise.

## FINDING

Reliable methods for prospective and retrospective evaluation of interdisciplinary research and education programs will require modification of the peer-review process to include researchers with interdisciplinary expertise in addition to researchers with expertise in the relevant disciplines.

**TOOLKIT**

**BOX 8-6 Determining How to Assess a Program: The Case of the Transdisciplinary Tobacco Use Research Centers**

The Transdisciplinary Tobacco Use Research Centers (TTURCs) were created in 1999 with funding from the National Cancer Institute, the National Institute on Drug Abuse, and the Robert Wood Johnson Foundation.[a] As with the NSF IGERT program (see Box 8-4), the funding agencies required that centers include a core program on evaluation. TTURC evaluation researchers have developed the following outcome metrics for measuring and evaluating science that bridges two or more disciplines:[b]

• How well is the *collaborative transdisciplinary work* of the centers (including *training*) accomplished?
• Does the collaborative transdisciplinary research of the centers lead to the development of new *or improved research methods* and/or *new or improved scientific models and theories*?
• Does research result in scientific publications that are *recognized* as high quality?
• Does research get *communicated* effectively?
• Are models and methods translated into *improved interventions*?
• Does research influence *health practices, health policy, or health outcomes*?

The evaluation focuses on the program as a whole and not necessarily on the individual research centers. To answer the questions, researchers analyze annual progress reports and the federal financial report and conduct a survey of each of the researchers involved. That is accomplished through survey analysis, content analysis of the progress report, peer evaluation of the progress report, bibliometric studies, peer evaluation of the publications, personnel analysis, and financial analysis. To date, data indicate progress toward intellectual integration within and between centers and changes in collaboration behaviors, and they highlight how pathways to integration are affected by environmental, organizational, and institutional factors.

---

[a]The Web Center for Social Research Methods. *http://www.socialresearchmethods.net/*. Accessed June 11, 2004.

[b]Stokols, D., et. al. "Evaluating Transdisciplinary Science," *Nicotine and Tobacco Research*. (2003) 5:1-19.

## RECOMMENDATIONS

**E-1: IDR programs and projects should be evaluated in such a way that there is an appropriate balance between criteria characteristic of IDR, such as contributions to creation of an emerging field and whether they lead to practical answers to societal questions, and traditional disciplinary criteria.**

For example, organizations that review IDR can measure

- The degree to which IDR contributes to the creation of an emerging field or discipline; emerging fields have included nanoscience and nanotechnology and cognitive science.
- How well IDR enhances the training of students and the careers of researchers in ways that surpass the results expected from disciplinary research; these might include employment in a broader array of positions, more rapid progress in gaining tenure and other goals, and greater numbers of speaking invitations.
- Whether the research leads to practical answers to societal questions; for example, an IDR effort to reduce hunger should produce some measurable progress toward that goal. The same IDR program might produce additional outcomes of value, including basic research, that were not expected.
- Whether participants demonstrate an expanded research vocabulary and abilities to work in more than one discipline.
- The extent to which IDR activities, institutes, or centers enhance the reputation of the host institutions; reputation can be measured in research funding, external recognition of IDR leadership, awards, and recognition of participants in the research.
- The long-term productivity of a program; not all initiatives will have the same lifetime, and the use of "sunset" provisions should be considered in the planning of IDR centers and programs.
- Multiple measures of research success, as appropriate to the fields being evaluated, such as conference presentations or patents in addition to publication in peer-reviewed journals.

**E-2: Interdisciplinary education and training programs should be evaluated according to criteria specifically relevant to interdisciplinary activities, such as number and mix of general student population participation and knowledge acquisition, in addition to the usual requirements of excellence in content and presentation.**

For example, organizations reviewing interdisciplinary education and training programs can begin with such criteria as the following, to be supplemented with others appropriate to the organizations' missions:

- Are interdisciplinary courses attracting more of the general student population to science and engineering courses?
- Are interdisciplinary courses and programs attracting a new or different mix of students to careers in science and engineering?
- Are interdisciplinary courses effective in instilling scientific and technologic literacy and awareness of the roles of science and technology in modern life?

**E-3: Funding organizations should enhance their proposal-review mechanisms so as to ensure appropriate breadth and depth of expertise in the review of proposals for interdisciplinary research, education, and training activities.**

For example, organizations that fund IDR could

- Involve researchers who have experience with and are knowledgeable about interdisciplinarity and ensure representation of the most important disciplinary points of view on panels that review IDR proposals.
- Evaluate a proposal to its cell-biology research program by using researchers in cell biology and including a substantial number in chemistry, physics, computer science, the social sciences, and the humanities as appropriate; this practice would help to ensure disciplinary breadth and reduce bias.
- Review a proposed interdisciplinary program in climate change by using input not only from experts in climate change and related fields—such as oceanography, meteorology, atmospheric chemistry, and land use—but also experts in the constituent disciplines—such as physics, chemistry, and statistics—and nonresearchers for whom the research is relevant; the contributions of different disciplines might be submitted separately in written form, and these reports would be offered to a full review panel, including both disciplinary and interdisciplinary researchers, for final assessment.

**E-4: Comparative evaluations of research institutions, such as the National Academies' assessment of doctoral programs and activities that rank university departments, should include the contributions of interdisciplinary activities that involve more than one department (even if it involves double-counting), as well as single-department contributions.**

For example, organizations that evaluate such institutions can

- Survey emerging interdisciplinary fields to identify demographic information (e.g. numbers and characteristics of participants in various interdisciplinary fields, and/or the the kinds of activities in which they are engaged).
- Experiment with "matrix evaluation" to capture the activities and

accomplishments of interdisciplinary researchers; a matrix approach is one that would consider IDR as an integral part of the disciplines in which the researchers are "embedded" and make visible the cross-departmental efforts of the researchers who make up the interdisciplinary teams.

• Include as evaluation criteria the comentoring of doctoral students, the contributions of individuals to multiple departments, and publication criteria. The publication criteria might include the nature of the journal audiences for whom the work is published; citation analysis that reveals a broad interdisciplinary interest in the work being cited; "double counting" of publications, by which credit for a given paper is awarded to all coauthors; multiple-authorship patterns that would reveal the disciplinary backgrounds of coauthors; and others that are still being developed.

• Include the facilitation of interdisciplinarity as part of the accreditation process.

# 9

# Toward New
# Interdisciplinary Structures

Performing interdisciplinary research (IDR) often requires additional resources, such as extra startup time, complex equipment, and extended funding. The nature of the structure in which IDR takes place—which may be an actual or virtual space—can help or hinder its progress. The hindrances created by some structures, discussed in Chapters 4 and 5, have prompted experiments designed to lower and even remove barriers and to facilitate IDR in other ways.

---

**Convocation Quote**

The academic research community has yet to grasp completely the degree to which interdisciplinary research probes at the heart of what the American research system has come to be, at least in terms of the role of the independent investigator. There are deep cultural issues for individual researchers and the institutions where they do their work that are so embedded it is going to take a lot of work to overturn them. It is not going to happen very easily.

James Collins, Arizona State University

---

## INTERDISCIPLINARY STRUCTURES

Over the last several decades, a variety of formats or structures for IDR have evolved. If they could be arranged along a spectrum, at one end might be the individual researcher—the modern equivalent of the polymath—who has achieved single-handedly a deep understanding of two or more disciplines and the ability to integrate them. At the other end might be a structure of multiple government-funded programs staffed by thousands of scientists and engineers drawn to a goal as ambitious—and focused—as the search for life on Mars. Interdisciplinary structures may also be interinstitutional, sharing no common physical space, or they may be in physical centers or "collaboratories" of substantial size and life span.

Whatever their structure, interdisciplinary projects flourish in an environment that allows researchers to communicate, share ideas, and collaborate across disciplines. The flow of ideas and people is made possible by institutional policies that govern faculty appointments and salary lines, faculty recruitment, responsibility for tenure and promotion decisions, allocations of indirect-cost returns on grants, development of new course and curricular materials, and so on.

## A VISION OF NEW INSTITUTIONAL STRUCTURES: THE MATRIX MODEL

Many researchers want to pursue interdisciplinary work more actively, but what new structures can best support them? The committee envisioned two possible modes for the creation of new structures: an incremental mode, which builds on lessons learned in the recent past, and a more transformative mode in which change comes more rapidly and discontinuously with respect to existing structures and practices.

Given the diverse nature of interdisciplinary activities, the number of formats for IDR in the future is likely to reflect the growing complexity of research. Whatever format characterizes a given IDR project, especially in academic institutions, it must operate in the context of a larger, overarching institutional framework that in many ways defines and constrains it. It is important, therefore, to examine institutional organizations and traditions critically and to ask what kinds of changes are possible and helpful for IDR.

An older management structure of universities is a landscape of separate components, or "silos," with weak coupling between them. A newer structure, which can already be discerned both in the United States and abroad and which has long been evident in industry and elsewhere, is more like a matrix, in which people move freely among disciplinary departments that are bridged and linked by interdisciplinary centers, offices, programs, courses, and curricula. There are many possible forms of coupling between departments and centers, including appointments, salary lines, distribution of indirect-cost returns, teaching assignments and course-teaching credits, curricula, and degree-granting.

A matrix structure (see Box 9-1) in a university might include many

---

DEFINITION

## BOX 9-1    What Is Matrix Management?

Matrix management is a product of organizational theory. The term refers to a management approach that encourages the development of orthogonal (cross-cutting) organizational structures. Traditionally, the department is the primary organizational structure of a university. Departments may be considered "vertical" structures. Orthogonal structures are functional groups that involve members who span multiple departments.

Some institutions have adopted matrix structures in which colleges, departments, and professional schools form the vertical dimension and research centers and institutes constitute the orthogonal dimension. In this spirit, the National Science Foundation (NSF) has established a suite of cross-cutting programs that include interdisciplinary programs, programs that are supported by multiple NSF directorates, and programs jointly supported by NSF and other federal agencies. The University of California, Davis has established horizontal budgeting structures (see Box 5-8). The University of Kansas has developed a matrix whereby research centers and institutes' directors report to the same central research administration as the departments. Benefits of this matrix structure include pooling of resources for equipment, grant-management support, generation of "critical mass," enhancement of stature, and mentoring, all of which improve the productivity of research faculty members.

---

[a]National Science Foundation Crosscutting/Interdisciplinary Programs home page *http://www.nsf.gov/home/crssprgm/*.

[b]Roberts, J. A. and Barnhill, R. E. "Engineering Togetherness: An Incentive System for Interdisciplinary Research." ASEE/IEEE Frontiers in Education Conference, Reno, NV, October 10-13, 2001. The authors write, "This type of organization, when properly implemented, facilitates interdisciplinary research.... Universities that tie research centers and institutes to disciplinary academic units will increasingly find themselves at a disadvantage in attempting to form effective teams to compete for interdisciplinary research grants which are more and more becoming the norm." See also Barnhill, R. E., "How sustainable is the modern research university." AAAS S&T Policy Forum. Washington, DC, April 23, 2004 *http://www.aaas.org/spp/rd/barnhill404.pdf.*

---

joint faculty appointments and PhDs granted in more than one department which would enable participants to address cross-cutting questions more easily. It might create numerous interdisciplinary courses for undergraduates, provide mentors who bridge the pertinent disciplines, and, equally important, offer faculty numerous opportunities for continuing education whereby they could add both depth and breadth of knowledge throughout their careers.

Successful matrix structures in research universities of the future may provide robust mechanisms for allocating faculty positions to areas of IDR, cross-departmental mechanisms for tenure and promotion review, and ways to facilitate team teaching by more flexible allocation of instructional credits. Policies that allow the return of some indirect-cost revenues to research units can be structured so as not to disadvantage interdisciplinary centers and programs that have external funding. Support for graduate students who choose to study in cross-disciplinary fields with mentoring by more than a single faculty member can create incentives for venturing into IDR. Most of those institutional changes would probably involve little cost; rather, they represent a shifting of existing incentive structures.

In the United States, many universities and other institutions are experimenting with matrix-like structures. At the US Geological Survey, researchers work in teams, but their funding may come from various programs not directly related to the teams. At the University of Washington, the Program on the Environment (PoE) has created a horizontal network to bring together faculty and students from across the university to participate in the environmental education programs (see Box 9-5). The PoE is overseen by a Governing Board that consists of 24 faculty, staff, and students representing a wide array of departments, colleges, and service units. In addition to an interdisciplinary bachelor's degree program, the PoE offers graduate certificates in three interdisciplinary fields.

## BEYOND THE MATRIX

Individual students, postdoctoral scholars, faculty, staff, and other members of academic communities accommodate their aspirations and plans to the possibilities that they see in the institutional structures around them. In considering how institutional characteristics might be changed to facilitate IDR, it is useful to think of how such changes might affect peoples' abilities to reach their goals. A more dramatic or "revolutionary" vision of interdisciplinarity might be seen as a transformed matrix in which institutions strive for a more complete integration of disciplines, institutions "without walls," a high degree of flexibility and mobility for students and faculty, and research efforts that are organized around problems rather than disciplines.

An example of a "revolutionary" vision is one in which students are encouraged to look across and draw experience from a wide spectrum of scientific knowledge and mentors before choosing a field of specialization (see Box 9-2). Some graduate programs, for example, admit students into the general "biological sciences" and allow them a year or two to choose a specialization. Similarly, the new Olin College, recently founded in Cambridge, Massachusetts, trains its incoming students simultaneously in all the engineering sciences; as students gain experience, they choose specific problems to focus on; in this case, the Olin Foundation has decided to pay all student tuition and to support the college itself for a specified period.

## BOX 9-2 Replacing Courses and Majors with Programs and Planning Units

In the traditional academic term, students take several stand-alone courses offered by individual departments and integrate cross-cutting concepts on their own. In contrast, students at The Evergreen State College[a] are strongly encouraged to take a single *program* each term. Programs, taught by faculty teams, are designed to help students to bring together ideas from multiple disciplines, with titles such as "Leadership for Urban Sustainability," "Fishes, Frogs, and Forests," and "Data to Information: Computer Science and Mathematics." Programs are organized into "planning units" associated with faculty who have related interests. Planning units offer students a means of focusing their study; students at Evergreen end up receiving a BA or a BS without a listed major. Graduate programs are similarly organized. For example, the Graduate Program in Environmental Studies was established in 1984 and integrates the study of environmental science and public policy. The curriculum consists of closely integrated courses taught by faculty teams trained in the social, biological, and physical sciences.

Other universities have adopted similar models. Pennsylvania State University offers intercollege programs for undergraduate minors in astrobiology, environmental inquiry, gerontology, marine sciences, military studies, and neuroscience.[b] The Department of Physics at Harvard has offered a joint concentration with the Department of Chemistry for many years.[c] The concentration in chemistry and physics is supervised by a committee that comprises members of the Departments of Physics and Chemistry, and it is administered through the office of the director of undergraduate studies. As the name implies, the concentration has been established to serve students who want to develop a strong foundation in both physics and chemistry rather than specialize in one or the other. The concentration is often chosen by students whose career goals lie in medicine, but the intellectual disciplines involved provide a suitable background for careers in a variety of professions. Some 15 years ago, 14 students opted for this honors program; over the years, enrollment has steadily increased, and in 2004 there are 45 students.

[a]Evergreen State College home page. *http://www.evergreen.edu.*
[b]Penn State University Intercollege Program home page *http://www.psu.edu/bulletins/ bluebook/$inmenu.htm.*
[c]Harvard University Chemistry and Physics Concentration home page *http://www. registrar.fas.harvard.edu/handbooks/student/chapter3/chemistry_and_physics.html* 2004.

There are models of interdisciplinarity in all venues of scholarship. Rockefeller University is organized around its laboratories (see Box 9-3); the Institute for Advanced Study in Princeton, New Jersey, admits only postgraduate "visiting members" who are free to pursue independent study and develop collaborations as they choose. The Theory Group at Microsoft Corp. and some national laboratories have no disciplinary divisions.

INNOVATIVE PRACTICES

**BOX 9-3    A University Without Departments:**
**Rockefeller University**

The Rockefeller Institute/University in New York City has been the site of more major discoveries in biomedicine in the 20th century than any other institution in the world. Rockefeller has been associated with 23 Nobel laureates and 19 Lasker Award recipients. Five faculty members have been named MacArthur fellows and 12 have garnered the National Medal of Science, the highest science award given by the United States. In addition, 32 Rockefeller faculty are elected members of the National Academy of Sciences.

Hollingsworth and Hollingsworth[a] argue that "major discoveries occurred repeatedly because there was a high degree of interdisciplinary and integrated activity across diverse fields of science, and because of leadership that gave particular attention to the creation and maintenance of a nurturing environment, though with rigorous standards of scientific excellence." In essence, there are three important characteristics: a high level of scientific diversity, low levels of internal differentiation (i.e., no disciplinary departments), and visionary leadership.

The Rockefeller Institute was founded not on the basis a particular field or researcher, but to pursue diverse subjects in biomedical sciences. Researchers with diverse scientific and cultural backgrounds were recruited. Most worked in fields that crossed academic disciplines. In addition, Rockefeller did not organize the production of knowledge around academic disciplines. The institute was originally organized around two departments: the Department of Laboratories and the Department of the Hospital. The university's laboratory-based organizational structure "without walls" and pared-down layers of administration do away with the schools and academic departments that too often separate scientists. "This approach fosters a tremendously rich soup of interdisciplinary research and collaboration," says Rockefeller Professor and Nobel laureate Günter Blobel.[b]

---

[a]Hollingsworth, R. and Hollingsworth, C.T. Major Discoveries and Biomedical Research Organizations: Perspectives on Interdisciplinarity, Nurturing Leadership, and Integrated Structure and Cultures. In: *Practising Interdisciplinarity*. Eds. Weingart, P. and Stehr, N., Toronto: University of Toronto Press, 2000, pp. 215-44.

[b]Rockefeller University home page *http://www.rockefeller.edu/about.php*.

Some of the innovations and experiments stem from the growing literature showing that organizing information into a conceptual framework allows a student to apply what was learned in new situations and to learn related information more quickly.[1] For example, students may find that the essence of physics is best discovered by beginning with specific methods—by "learning how to learn"—rather than by beginning with formulas, facts,

---

[1]See for example, National Research Council "*How People Learn: Brain, Mind, Experience, and School.*" 2000. Washington, D.C.: National Academy Press.

and laws whose utility or relevance they can better appreciate at a later stage of education. Clearly, institutions that implement the kinds of changes described are placing a heavy burden of decision making on their students. The students in turn must rely on deeper and more extensive networking with teachers, mentors, and other students.

## SUPPORTING NEW INTERDISCIPLINARY STRUCTURES FOR PEOPLE AND PROGRAMS

Is it reasonable for institutions of higher learning to remake themselves around new interdisciplinary structures of teaching and research (see Box 9-4)? This committee has heard many arguments for change, as well as reasons for caution. Few voices, for example, have been raised in favor of

---

### INNOVATIVE PRACTICES

#### BOX 9-4   Cross-Cutting Reorganization of Academic Departments

Funding agencies can act as change agents, providing funding for programs to invigorate an emerging field of study or to establish new priorities for research universities. At least partly as a consequence of participating in an engineering research center (see Box 8-2), the Schools of Engineering of Purdue University in 2003 decided to reorganize into clusters, or signature areas. These areas are "multidisciplinary initiatives which cut across the established boundaries of Purdue's engineering schools and related disciplines."[a]

To support the eight newly created areas, Purdue is investing in new faculty positions that will be filled by using a cluster hiring process (see Box 5-4). Areas include advanced materials and manufacturing, global sustainable industrial systems, intelligent infrastructure systems, and nanotechnologies and nanophotonics. Purdue is also expanding and upgrading facilities, including the development of a transparent environment for multidisciplinary work.

The primary goal of the reorganization is to provide an opportunity for undergraduate and graduate engineering students to learn and work in an interdisciplinary environment and to gain real-world experience. To that end, Purdue has created a new Department of Engineering Education.[b] The new department will combine the existing freshman engineering and interdisciplinary engineering programs and aims to increase student interest in engineering and research in how students learn engineering concepts.

---

[a]Purdue University, College of Engineering, Signature Areas. *https://engineering.purdue.edu/Engr/Signature.*

[b]Holsapple, M. Purdue Counters Trend, Engineers Education from the Ground Up. Purdue News, April 9, 2004. Available on line at: *http://news.uns.purdue.edu/UNS/html3month/2004/040409.BOT.enged.html.*

abolishing sound institutional management that is needed to organize, support, and legitimize research programs. Many voices have confirmed the importance of mastering a specific discipline in depth before investigating new disciplines. And no one has pushed for institutional change that is forced or attempted in precipitous fashion. It seems more reasonable for institutions to adopt goals that look revolutionary now but to approach them in ways that are based on consensus, experiment, and sound models.

What might be some useful features of the restructured university—one that serves the interests of students, faculty, and the institution? The following suggestions are intended to put forward directions of desirable change without constituting recommendations. Most of these steps have been tested by institutions and might serve as models for others.

## Change at the Undergraduate Level

Undergraduate students might profit by planning programs that suit their interests and abilities with continual reshaping in the light of advancing understanding and with the guidance of faculty mentors. Graduation requirements could be general, including such broad features as total amount of coursework required and requirements for independent study or research. Focused interdisciplinary programs, such as those at the intersections of natural science and social science, could be taught by teams of interdisciplinary faculty working outside the aegis of individual departments or colleges. Students could be encouraged to become active members of interdisciplinary research groups and to adopt roles commensurate with their skills, talents, and goals.

Undergraduate students have shown themselves to be responsive to interdisciplinary and problem-driven questions, especially those of societal relevance. (See Figures 4-1 and 8-1.) They can prepare to address such questions by seeking institutions that provide opportunities for IDR at the undergraduate level, have strong interdepartmental connections and interdisciplinary centers and programs, provide opportunities for cooperative experiences outside academe, and allow dual or multiple majors or majors and minors in different fields.

## Change at the Graduate Level

Many institutions already admit graduate students to programs of study, some interdisciplinary, whose admissions criteria, degree requirements, and formation of graduate-study committees are administered through the programs themselves. Policies and practices are normally set by faculty members recruited into the programs. (See, for example, Boxes 4-2, 4-3, 4-4, and 4-5.)

In a more extensive implementation of this model, decisions about allocations of faculty positions to various programs, research budgets, and teaching budgets could be made by deans with responsibility for groups of programs. The graduate programs could place a premium on team teaching and on finding dual faculty mentors (see Box 4-5). Graduate degrees could be awarded by the programs with an optional focus on a particular discipline(s); for example, a student might receive a PhD in climate modeling with a focus in geology, atmospheric science, or chemistry.

Successful implementation of such a vision requires a matrix model in which the distribution of such important resources as research space and graduate-teaching-assistant positions is determined for the university as a whole rather than at a departmental or perhaps even college level.

## Change at the Faculty Level

Faculty could be recruited for positions in programs as well as in departments (see Box 9-5). They could teach courses within the special sphere of a program or foundation courses in traditional areas. Advancement toward tenure could be monitored by one or more mentors in the faculty member's program and by senior faculty in traditional fields of special interest to the young faculty member. Active participation and effectiveness in one or more program areas could be expected of all faculty seeking tenure. Membership in any specific program would probably not be permanent; the program might disappear or evolve, or the faculty member's interests might change.

The concept of tenure could be more flexible. Faculty admitted to tenure after initial evaluation—after, say, 5 or 7 years—might receive 5-year reappointments. Reappointment might depend on successful review by a peer faculty committee in the areas of specialization, including external

---

**INNOVATIVE PRACTICES**

**BOX 9-5  Cohiring: Collaborations between Centers and Departments[a]**

How do IDR centers, which generally operate without the ability to hire faculty or grant degrees independently, attract faculty and students? Joint appointments are one way of solving this dilemma; but in many cases, faculty report personal and departmental dissatisfaction at determining just how to apportion and credit percentages of time. Cohiring is an innovative method for bringing faculty into centers.

The University of Washington Program on the Environment[b] (PoE) is a horizontally organized universitywide institute. The PoE is not a traditional academic department and does not have a faculty of its own. Instead, it plays a networking role, bringing together faculty and students from across the university to augment existing programs and to offer integrated, interdisciplinary programs that cross traditional disciplinary boundaries. Instead of allocating faculty lines, the university president has set aside a permanent budget that the PoE uses to hire faculty in collaboration with departments and schools. By obligating a smaller fraction of the PoE operating budget, this enhances the flexibility and adaptability of the program and removes it from competition with departments and colleges. Cohiring enables the university to benefit from the presence of scholars who would not readily fit into pre-existing departmental frameworks. The PoE pays for a portion of the startup costs and salary for the first 3-5 years, after which the department becomes fully responsible for the faculty member. Colleges and departments are strongly encouraged to donate faculty time to the teaching of environmental-studies courses. Student credit-hours accruing from such teaching are credited to the faculty mem-

reviewers when appropriate. The principles of academic freedom that gave rise to the tradition of indefinite tenure can be protected by strong contractual agreements and the use of multiyear rolling appointments so that no faculty member would be subject to dismissal suddenly or without substantial cause.

The concept of the university professorship, in which the recipient is appointed "at large" and not to a specific department, which allows the recipient to move between departments, could be expanded without changing the nature of departments.

## Change at the Institutional Level

At the level of colleges and large institutions, the university could remain organized in more or less traditional fashion, including "colleges" of science, humanities, social sciences, engineering, education, and so on. How-

---

bers' home departments. The PoE can also use its budget to compensate departments for faculty teaching ("release time") in the program.[c]

The Center for the Neural Basis of Cognition (CNBC)[d] is a joint program of Carnegie Mellon University and the University of Pittsburgh. As in the PoE, the CNBC directors have spent a huge amount of time in building relationships with affiliated departments. Their overall goal is to make it clear that connections with other units are mutually beneficial: where disciplines can be seen as atoms of an inert gas, departments can bring people together with van de Waals forces, but the CNBC director says that "almost all members are in a covalent relationship." Faculty are hired collaboratively but appointed to a home department. Center funds are used to help with startup costs, and the departments thereafter assume responsibility for the hire. Promotion and tenure are integrated. Tenure decisions are made at the departmental level, but the center director is involved. Also bringing an interdisciplinary perspective to the review committees are the faculty associated with the center who are already tenured and serve on several departmental review committees.

---

[a]Partially derived from staff-conducted interviews with Ed Miles, chair of the Task Force on Environmental Education, and professor, School of Marine Affairs and Graduate School of Public Affairs (July 16, 2003); and James McClelland, codirector, Center for the Neural Basis of Cognition, Carnegie Mellon University (June 26, 2003).

[b]University of Washington, Program on the Environment home page http://depts.washington.edu/poeweb/about/index.html.

[c]For more on how appointments of faculty members are administered at the University of Washington, see http://www.washington.edu/tfee/final96.txt.

[d]Center for the Neural Basis of Cognition home page http://www.cnbc.cmu.edu/.

ever, these colleges could have much more porous boundaries than they do now. Faculty appointments could be more readily allocated and moved into and between colleges.

---

**Convocation Quote**

What we have found is that full-time long-term collaborations are actually not that effective. They reduce interaction, and they reduce innovation. What we need to think about is establishing long-term organizational structures that allow for short-term intensive collaboration experiences.

Diana Rhoten, Director, Hybrid Vigor Institute, and program officer, Social Science Research Council

---

For example:

• A faculty member with a JD degree who is interested in international law might have an appointment in a program that focuses on global hunger or on global technology transfer; the person might spend a year in team teaching in that program and the next year in teaching a foundation course in law, such as civil procedures.

• Space could be regarded as a fungible asset (see Box 9-6) so that hiring of a new faculty member in chemistry who requires wet-laboratory space might depend on arranging suitable laboratory space. The authority to make and budget for such space allocations could reside in the office of the dean or provost.

• Programs might lie not within the purview of colleges, but rather at a higher level, spanning more than one college. Furthermore, programs could be reviewed periodically, with the option of terminating those that no longer addressed subjects of high priority (see Box 5-6). The distribution of resources between colleges and programs might depend on the character of the institution, such as whether it is a private or publicly supported institution. The general objective would be to maintain a high degree of flexibility and to avoid a stultifying concentration of influence and authority at lower levels of organization.

## CHANGE DRIVEN BY GENERATIVE TECHNOLOGIES

Some technologies are changing not only how researchers work on their projects but also how they work with one another. For example, the sharing of information and even the development of ideas are assisted by new ways of communicating, manipulating, storing, retrieving, and analyzing information. More and more meetings are held by using "shared-

---

**INNOVATIVE PRACTICES**

**BOX 9-6  Hotel Space: The Allocation of Space by Project**

Stanford's Bio-X project is an ambitious initiative designed to facilitate IDR in subjects related to biology and medicine. The physical center of the project is its newly constructed Clark Center facility, home to about 40 faculty whose interests span the scientific disciplines. Each faculty member has the traditional associated laboratory space.

The Bio-X project is also experimenting with a new model for space allocation. Some 65 benches have been set aside for temporary occupancy and designated "hotel space."[a] The benches are designed to provide an opportunity for researchers to work in proximity during the early stages of projects, and occupancy is not to exceed 12 months. Hotel space is allocated by the Bio-X Leadership Council, which is a faculty group charged with planning the Bio-X program.

The Clark Center is still in its early stages of operation, but hotel space is intended to stimulate collaboration by encouraging scientists and engineers in disparate disciplines to work together. Visiting researchers may have a specific vision for collaborating with Clark Center researchers or other visiting researchers. Other visitors may simply want to work next to researchers doing a particular type of work to investigate the possibility of collaboration.

The Bio-X project views hotel space as an experiment unto itself, but this will not be the only experiment of its kind. A similar approach is planned for the Janelia Farms research campus of the Howard Hughes Medical Institute (see Box 6-7). Hotel space is one of several revolutionary approaches whose value will become clearer as interdisciplinary projects mature.

---

[a]Stanford University Bio-X, Hotel Space in the Clark Center *http://biox.stanford.edu/clark/hotel_info1.html.*

---

whiteboard" software that allows participants to conduct virtual meetings; display drawings, slides, or equations; compose a document together; and poll participants instantly. Many traditional researchers insist on the need for face-to-face meetings to forge effective collaborations, but younger people growing up in a world of instant messaging may develop virtual modes of collaboration that are equally or even more effective.

Information technologies are already generating powerful new cyberstructures. For example, new techniques have made possible the design and implementation of the National Institutes of Health Biomedical Informatics Research Network (BIRN), which uses a distributed information technology infrastructure to coordinate biomedical research in multiple institutions (see Box 9-7). In what BIRN calls its "evolving cyberinfrastructure," a coordinating center was established in 2001 to achieve large-scale data-sharing among far-flung "test beds" working with brain morphometry (six

---

INNOVATIVE PRACTICES

## BOX 9-7  Supporting Teamwork with Distributed Information Technologies: The Biomedical Informatics Research Network (BIRN)

As the amount, size, and complexity of data increase, the finding and extraction of relevant information by individual scientists become more difficult. But amid the growing complexity are unprecedented opportunities for data-sharing and data-mining. Cyberinfrastructures, also known as grids, can create structured database repositories that facilitate data accessibility and foster collaboration. The Biomedical Informatics Research Network (BIRN) is one such grid project.[a] BIRN is supported by the National Centers for Research Resources of the National Institutes of Health. Its goal is to establish an information technology infrastructure to enable fundamentally new capabilities in large-scale studies of human disease. BIRN involves a national consortium of 12 universities and 16 research groups. It consists of three test-bed projects that are conducting structural and functional studies of neurological disease: Function BIRN, Morphometry BIRN, and Mouse BIRN.

A central premise of BIRN is that the location of data and resources is less important than their organization and accessibility. One of BIRN's core efforts is to develop technologies to ensure that each BIRN site and test bed can create and manage sophisticated and highly structured data repositories. To that end, a coordinating center (CC) was established in 2001 to develop, implement, and support the information infrastructure necessary to achieve large-scale data-sharing among participants.

BIRN-CC is a partnership of computer scientists, neuroscientists, and engineers who as equal partners address a large variety of technical, policy, and architectural issues.[b] The collaboration is truly interdisciplinary, inasmuch as CC members must be interested in and committed to learning each other's disciplinary language so that they can work effectively toward common goals. In addition to designing infrastructure, the BIRN-CC is responsible for encouraging interactions among BIRN participants: the CC manages the BIRN Web site and newsletter and organizes an annual meeting to define collaborative needs and set research priorities.[c]

---

[a]Biomedical Informatics Research Network home page. *http://www.nbirn.net.*

[b]Lin, A. W., Maas, P., Peltier, S., Ellisman, M. (2004) Harnessing the Power of the Globus Toolkit. *Cluster World.* 2(1):12-14, 54.

[c]James, M. (2004) Productive All Hands Meeting Defines CC Goals. *BIRNing Issues.* 2(2):10.

---

institutions), schizophrenia (11 institutions), and mouse models of neurological disorders (four institutions).

## CONCLUSIONS

As interdisciplinary research, scholarship, and teaching increase in importance in institutions of higher education, so does the urgency to find new policies and structures that accommodate interdisciplinarity. Successful institutions are likely to be those that are nimble and willing enough to develop such policies. The likely outcomes of the policies could be higher levels of external support for the institutions, greater success in recruiting the most promising new faculty and students, and enhanced service to society in the form of successful scholarship and research at the frontiers of knowledge.

## FINDING

The increasing specialization and cross-fertilizations in science and engineering require new modes of organization and a modified reward structure to facilitate interdisciplinary interactions.

## RECOMMENDATIONS

**U-1: Institutions should explore alternative administrative structures and business models that facilitate IDR across traditional organizational structures.**

For example, institutions can

- Experiment with alternative administrative structures, such as the matrix model, in which people move freely among disciplinary departments that are bridged and linked by interdisciplinary centers, offices, programs, and curricula or, alternatively, create institutions "without walls" that have no disciplinary departments and are organized around problems rather than disciplines.
- Facilitate the offering of multidisciplinary courses, provide graduate students with multiple mentors, and offer faculty numerous opportunities for continuing education.
- Oversee interdisciplinary programs at the university level rather than that of a single college.
- Review programs periodically with the option of terminating those no longer of high priority so that there is flexibility to respond to emerging opportunities.

**U-2: Allocations of resources from high-level administration to interdisciplinary units, to further their formation and continued operation, should be considered in addition to resource allocations of discipline-driven departments and colleges. Such allocations should be driven by the inherent intellectual values of the research and by the promise of IDR in addressing urgent societal problems.**

For example, institutions can

- Put in place policies that allow the return of some indirect cost revenues to research units such that interdisciplinary centers and programs with external funding are not disadvantaged.
- Provide support for graduate students who choose to study inter-disciplinary fields with mentoring by more than a single faculty member.
- Provide support for generative technologies that allow the sharing of information and ideas.
- Invest federal funds in activities that lead to the design and implementation of research activities that take full advantage of a distributed information technology infrastructure to coordinate research across institutional lines.

**U-3: Recruitment practices, from recruitment of graduate students to hiring of faculty, should be revised to include recruitment across department and college lines.**

For example, institutions can

- Admit graduate students into broad fields (for example, biological sciences as opposed to microbiology; engineering as opposed to mechanical engineering) with no requirement to specialize until the end of the first or second year.
- Increase the number of joint faculty appointments and PhD programs from a few to many.
- Recruit faculty for positions both in programs and in departments so they can teach both within the special sphere of a program and in foundation courses in traditional areas.

**U-4: The traditional practices and norms in hiring of faculty and in making tenure decisions should be revised to take into account more fully the values inherent in IDR activities.**

For example, institutions can

- Provide robust mechanisms for allocating faculty positions to areas of IDR.
- Provide cross-departmental mechanisms for tenure and promotion review.
- Monitor a tenure-track faculty member's progress toward tenure with both mentors from the faculty member's program and senior faculty in traditional fields of special interest to that faculty member.

U-5: Continuing social science, humanities, and information-science-based studies of the complex social and intellectual processes that make for successful IDR are needed to deepen the understanding of these processes and to enhance the prospects for the creation and management of successful programs in specific fields and local institutions.

# 10

# Findings and Recommendations

This chapter contains the committee's findings and recommendations, which have been gathered here from the foregoing chapters.

## FINDINGS

### Definition

1. Interdisciplinary research (IDR) is a mode of research by teams or individuals that integrates information, data, techniques, tools, perspectives, concepts, and/or theories from two or more disciplines or bodies of specialized knowledge to advance fundamental understanding or to solve problems whose solutions are beyond the scope of a single discipline or area of research practice.

### Current Situation

2. IDR is pluralistic in method and focus. It may be conducted by individuals or groups and may be driven by scientific curiosity or practical needs.

3. Interdisciplinary thinking is rapidly becoming an integral feature of research as a result of four powerful "drivers": the inherent complexity of nature and society, the desire to explore problems and questions that are not confined to a single discipline, the need to solve societal problems, and the power of new technologies.

At a variety of academic institutions, the number of departments has increased steadily over the last century, from about 20 in 1900 to between 50 and 110 in 2000.[1] National professional societies have also increased in number from 82 in 1900 to 367 in 1985.[2] Although those changes may appear to indicate increasing specialization, the increase in new departments and societies primarily reflects a blending of previously distinct fields to produce new areas such as biophysics and biochemistry and, more recently, neuroscience and photonics.

**4. Successful interdisciplinary researchers have found ways to integrate and synthesize disciplinary depth with breadth of interests, visions, and skills.**

Studies of expertise have shown that perception and understanding of a given task or problem depends on the knowledge a person brings to a situation.[3] A challenge in interdisciplinary work is to develop expertise in more than one area. Among the respondents to the committee's survey, 94% of whom were at least partially involved in IDR, clear strategies to obtaining discipline-spanning expertise emerged. Over half indicated that after developing expertise in one field, they had sought training in additional fields through postdoctoral fellowships, additional advanced degrees, or day-to-day interactions with researchers in different fields to participate in interdisciplinary projects. These strategies were reflected in the top recommendations respondents made for institutions, principal investigators, postdocs and students: foster a collaborative environment (26 percent), build a network with other researchers (20 percent), find a postdoctoral appointment in a different field (13 percent), seek additional mentors (12 percent), cross boundaries between fields (25 percent) and at the same time develop a solid background in one discipline (12 percent).

**5. Students, especially undergraduates, are strongly attracted to interdisciplinary courses, especially those of societal relevance.**

For example, at Harvard University, the number of undergraduate joint concentrations in chemistry and physics has risen from 14 to 45 over the last 15 years. There has been large-scale growth at Columbia College since 1993 in majors and concentrations in interdisciplinary departments and interdepartmental programs. At Stanford University, a multiyear decline in the number of students majoring in earth science was reversed when the

---

[1]See Figure 1-1.

[2]See Figure 7-1.

[3]National Research Council. How People Learn. Brain, Mind, Experience, and School. Washington, D.C.: National Academy Press. 2000.

major was changed from the single discipline of geology to the interdisciplinary "earth systems."

6. The success of IDR groups depends on institutional commitment and research leadership. Leaders with clear vision and effective communication and team-building skills can catalyze the integration of disciplines.

In the committee's survey, the top recommendation for principal investigators was to lead research teams in a way that is supportive of IDR (44 percent). Respondents also recommended that departments develop new organizational approaches permissive to IDR (32 percent).

### Challenges to Overcome

7. The characteristics of IDR pose special challenges for funding organizations that wish to support it. IDR is typically collaborative and involves people of disparate backgrounds. Thus, it may take extra time for building consensus and for learning of methods, languages, and cultures.

In the committee's survey, researchers' top three recommendations for institutions, project leaders, principal investigators, educators, postdoctoral scholars, and students focused on enhancing communication between researchers. Over 20 percent of the respondents stated specifically that principal investigators and postdocs need time to develop effective networks and research strategies.

8. Social-science research has not yet fully elucidated the complex social and intellectual processes that make for successful IDR. A deeper understanding of these processes will further enhance the prospects for creation and management of successful IDR programs.

### Changes Needed

9. In attempting to balance the strengthening of disciplines and the pursuit of interdisciplinary research, education, and training, many institutions are impeded by traditions and policies that govern hiring, promotion, tenure, and resource allocation.

In the committee's informal survey of those attending its workshop on IDR, 72 percent of respondents reported impediments to IDR at their institutions. Among researchers, the most common were promotion and tenure criteria (18 percent) and budget control (16 percent). Among provosts responding to the survey, the most common impediments were tenure and promotion criteria (19 percent) and space allocation (19 percent).

**10. The increasing specialization and cross-fertilization in science and engineering require new modes of organization and a modified reward structure to facilitate interdisciplinary interactions.**

In the committee's survey, the top recommendation for academic departments was to create and emphasize new organizational approaches, such as (1) hiring strategies and practices and (2) physical and personnel networks conducive to interdisciplinary exchange.

**11. Professional societies have the opportunity to facilitate IDR by producing state-of-the-art reports on recent research developments and on curriculum, assessment, and accreditation methods; enhancing personal interactions; building partnerships among societies; publishing interdisciplinary journals and special editions of disciplinary journals and promoting mutual understanding of disciplinary methods, languages, and cultures.**

**12. Reliable methods for prospective and retrospective evaluation of interdisciplinary research and education programs will require modification of the peer-review process to include researchers with interdisciplinary expertise in addition to researchers with expertise in the relevant disciplines.**

**13. Industrial and national laboratories have long experience in supporting IDR. Unlike universities, industry and national laboratories organize by the problems they wish their research enterprise to address. As problems come and go, so does the design of the organization.**

**14. Although research management in industrial and government settings tends to be more "top-down" than it is at universities, some of its lessons may be profitably incorporated into universities' IDR strategies.**

**15. Collaborative interdisciplinary research partnerships among universities, industry, and government have increased and diversified rapidly. Although such partnerships still face significant barriers, well-documented studies provide strong evidence of both their research benefits and their effectiveness in bringing together diverse cultures.**

## RECOMMENDATIONS

The committee's recommendations are listed here primarily by category of stakeholders in interdisciplinary research and education. The recommendations are based on the committee's deliberations and suggestions from convocation participants (students, researchers, academic and non-academic institutional leaders, funding organizations, and professional societies) the focus groups held at the Keck Futures Symposium, interviews with leading scholars, and responses to committee's surveys.

In each of the committee's category of pertinent stakeholders, general recommendations are presented in bold-face type. These are intended as guidelines or objectives derived from the findings of this study, and they are followed by suggestions of ways to implement them. The suggested actions, based on real examples and experiments summarized in the foregoing chapters, are intended not as prescriptions but as "templates" for people or organizations to adapt according to their particular situations and the availability of resources. The relatively large number of these actions is intended to indicate the diversity of possibilities.

The committee hopes that these templates will help people and organizations to design their own strategies, whether they are ready to act now or after months or even years of study and fund-raising. The committee also hopes that the results of this study provide convincing evidence both of the value of interdisciplinarity and of the urgent need to revise some traditions in academic, funding, and professional organizations in order to promote IDR.

### Students

**S-1:** ***Undergraduate students*** **should seek out interdisciplinary experiences, such as courses at the interfaces of traditional disciplines that address basic research problems, interdisciplinary courses that address societal problems, and research experiences that span more than one traditional discipline.**

For example, students can

• Begin preparation for IDR through an IDR project or summer IDR experience.

• Approach interdisciplinarity by first gaining a solid foundation in one discipline and then adding disciplines as needed. Additional courses provide opportunities to understand the culture of other disciplines, gain new skills and techniques, and network with other researchers.

**S-2:** ***Graduate students*** **should explore ways to broaden their experience by gaining "requisite" knowledge in one or more fields in addition to their primary field.**

For example, graduate students can

- Do this through master's theses or PhD dissertations that involve multiple advisers in different disciplines.
- Share an office with students in other fields.
- Enhance their interdisciplinary expertise by participating in conferences outside their fields and in poster sessions that represent multiple disciplines. Those venues provide opportunities for junior researchers to present their work to colleagues outside their fields.

### Postdoctoral Scholars

**P-1: Postdoctoral scholars can actively exploit formal and informal means of gaining interdisciplinary experiences during their postdoctoral appointments through such mechanisms as networking events and internships in industrial and nonacademic settings.**

For example, postdoctoral scholars can

- Seek formal and informal opportunities to communicate with potential research collaborators in other disciplines and develop a network of interdisciplinary colleagues.
- Broaden their perspective through internships in industrial settings or other nonacademic settings.

**P-2: Postdoctoral scholars interested in interdisciplinary work should seek to identify institutions and mentors favorable to IDR.**

For example, postdoctoral scholars can seek positions at institutions that

- Have strong interdisciplinary programs or institutes.
- Have a history of encouraging mentoring relationships across departmental lines.
- Offer technologies, facilities, or instrumentation that further one's ability to do IDR.
- Have researchers and faculty members with whom the postdoctoral scholar interacts place a high priority on shared interdisciplinary activities.

### Researchers and Faculty Members

**R-1: Researchers and faculty members desiring to work on interdisciplinary research, education, and training projects should immerse themselves in the languages, cultures, and knowledge of their collaborators in IDR.**

For example, researchers and faculty members can

- Develop relationships with colleagues in other disciplines.
- Learn more about the knowledge and culture of other disciplines by participating in interdisciplinary projects.
- Actively seek opportunities to teach classes in other departments and give papers at conferences outside their own disciplines or departments. In their written and oral communications, researchers and faculty members can facilitate IDR by using language that those in other disciplines are able to understand.
- Mentor students and postdoctoral scholars who wish to work on interdisciplinary problems.

**R-2: Researchers and faculty members who hire postdoctoral scholars from other fields should assume the responsibility for educating them in the new specialties and become acquainted with the postdoctoral scholars' knowledge and techniques.**

For example, researchers and faculty members can

- Familiarize themselves with the research cultures and evaluation methods of the postdoctoral scholars' fields.
- Learn about the career expectations of the postdoctoral scholars, when possible, and the demands that they will encounter in their careers.
- Guide the postdoctoral scholars toward interdisciplinary learning opportunities, including workshops, research presentations, and social gatherings.

### Educators

**A-1: Educators should facilitate IDR by providing educational and training opportunities for undergraduates, graduate students, and postdoctoral scholars, such as relating foundation courses, data gathering and analysis, and research activities to other fields of study and to society at large.**

For example, educators can

- Provide training opportunities that involve research, data-gathering, data analysis, and interactions among students in different fields.
- Demonstrate the power of interdisciplinarity by inviting IDR speakers, providing examples of major discoveries made through IDR, and highlighting exciting current research at the interfaces of fields.

- Encourage a multifaceted, broadly analytical approach to problem-solving.
- Include as part of foundation courses (such as general chemistry) materials that show how the subjects are related to other fields of study and to society at large.
- Show through explanatory examples the relevance of IDR to complex societal problems, which often require multiple disciplines and challenge current scientific and technical methods.
- Discourage the notion that some disciplines rank higher than others.
- Create more opportunities for students to learn how research disciplines complement one another by

— Developing policies and practices that support team teaching of interdisciplinary courses by faculty members in diverse departments or colleges.

— Modifying core course requirements so that students have more opportunities to add breadth to their study programs.

— Provide team-building and leadership-skills development as a formal part of the educational process.

## Academic Institutions' Policies

**I-1: Academic institutions should develop new and strengthen existing policies and practices that lower or remove barriers to interdisciplinary research and scholarship, including developing joint programs with industry and government and nongovernment organizations.**

For example, institutions can

- Provide more flexibility in promotion and tenure procedures, recognizing that the contributions of a person in IDR may need to be evaluated differently from those of a person in a single-discipline project. Institutions could

— Establish interdisciplinary review committees to evaluate faculty who are conducting IDR.

— Extend the venue for tenure review of interdisciplinary scholars beyond the department.

— Increase recognition of co-principal investigators' research activities during promotion and tenure decisions.

— Develop mechanisms to evaluate the contribution of each member of an IDR team.

- Establish institutional advisory committees of researchers successful in IDR to evaluate new proposals prior to implementation.

- Require regular reviews of IDR centers and institutes and establish sunset provisions, where appropriate, when they are initiated.
- Give high priority to recruitment of appropriate faculty and other researchers whose focus is interdisciplinary; this can be accomplished in part by allocating substantial resources to centrally funded, multidepartmental hiring of faculty and postdoctoral scholars and admission of graduate students.
- Coordinate hiring across departments and centers to maximize collaborative research and teaching possibilities.
- Develop joint IDR programs and internships with industry.
- Allow for the longer startup time required by some IDR programs.
- Gather information about the extent, quality, and importance of IDR in the institution and make the information available to faculty.
- Provide mechanisms to build a community of interdisciplinary scholars across the institution similar to the community that is in a department.

**I-2: Beyond the measures suggested in I-1, institutions should experiment with more innovative policies and structures to facilitate IDR, making appropriate use of lessons learned from the performance of IDR in industrial and national laboratories.**

For example, institutions can

- Experiment with alternatives to departmental tenure through new modes of employment, retention, and promotion.
- Selectively apply pooled faculty lines and funds available for startup costs for new faculty toward recruitment of faculty with interdisciplinary interests and credentials.
- Experiment with administrative structures that lower administrative and funding walls between departments and other kinds of academic units.
- Create laboratory facilities with reassignable spaces and equipment for people performing IDR.
- Create specific IDR grants and training programs for distinct career stages to assist in learning new disciplines and participating in IDR programs.
- Create mechanisms to fund graduate students and postdoctoral scholars whose research draws on multiple fields and may not be considered central to any one department.
- Develop a process for dealing with intellectual-property allocation that is consistent with encouraging IDR.
- Increase "porosity" across organizational boundaries by

— Encouraging joint recruitment and appointment of faculty through resources available centrally.

— Creating competitive internal leave for study in a new discipline, allowing faculty to take courses, training, and additional advanced degrees in their own universities.

— Encouraging departments and colleges to work with IDR centers and institutes in hiring faculty with interdisciplinary backgrounds.

— Providing fellowships that are portable within the institution.

— Allowing courtesy appointments that recognize interactions and collaborations across departments but that do not have the formal split responsibility of a joint appointment.

— Placing departments near one another to take advantage of their potential for fruitful interdisciplinary collaborations.

**I-3: Institutions should support interdisciplinary education and training for students, postdoctoral scholars, researchers, and faculty by providing such mechanisms as undergraduate research opportunities, faculty team-teaching credit, and IDR management training.**

Such education and training could cover interdisciplinary research techniques, interdisciplinary team management skills, methods for teaching non-majors, etc. For example, institutions can

• Provide more opportunities for undergraduate research experiences.

• Allow faculty to receive full credit for team teaching in interdisciplinary courses.

• Encourage multiple mentors for students and pairing of appropriate senior interdisciplinary faculty with junior faculty interested in IDR.

• Provide opportunities (such as sabbaticals) for students and faculty members to learn the content, languages, and cultures of disciplines other than their own, both within and outside their home institution.

• Support formal programs on the management of IDR programs, including leadership and team-forming activities.

**I-4: Institutions should develop equitable and flexible budgetary and cost-sharing policies that support IDR.**

For example, institutions can

• Streamline fair and equitable budgeting procedures across department or school lines to allocate resources to interdisciplinary units outside the departments or schools.

• Create a campuswide inventory of equipment to enhance sharing and underwrite centralized equipment and instrument facilities for use by IDR projects and by multiple disciplines.

- Credit a percentage of a project's indirect cost to support the infra-structure of research activities that cross departmental and school boundaries.
- Allocate research space to projects, as well as departments.
- Deploy a substantial fraction of flexible resources—such as seed money, support staff, and space—in support of IDR.

## Team Leaders

**T-1: To facilitate the work of an IDR team, its leaders should bring together potential research collaborators early in the process and work toward agreement on key issues.**

For example, team leaders can

- Catalyze the skillful design of research plans and the integration of knowledge and skills in multiple disciplines, rather than "stapling together" similar or overlapping proposals.
- Establish early agreements on research methods, goals and time-lines, and regular meetings.

**T-2: IDR leaders should seek to ensure that each participant strikes an appropriate balance between leading and following and between contributing to and benefiting from the efforts of the team.**

For example, leaders can

- Help the team to decide who will take responsibility for each por-tion of the research plan.
- Encourage participants to develop appropriate ways to share credit, including authorship credit, for the achievements of the team.
- Acquaint students with literature on integration and collaboration.
- Provide adequate time for mutual learning.

## Funding Organizations

**F-1: Funding organizations should recognize and take into consider-ation in their programs and processes the unique challenges faced by IDR with respect to risk, organizational mode, and time.**

For example, funding organizations can seek to

- Ensure that a request for proposals does not inadvertently favor funding a single-discipline project over an IDR project; for example, by including limitations on funding amounts, duration of funding (successful

IDR teams often take longer to build and to coalesce), scope, and allowable travel and other budget items, all of which would militate against IDR.

• Develop funding programs specifically designed for IDR, for example, by focusing research around problems rather than disciplines.

• Provide seed-funding opportunities for proof-of-concept work that allows researchers in different disciplines to develop joint research plans and to perform initial data collection or for new organizational models or project approaches that enable IDR.

• Have support for universities to provide shared research buildings, large equipment, or specialized personnel (machinists, glassblowers, and computer and electronic technicians).

• Provide funding mechanisms that allow researchers to obtain training in new fields.

• Fund programs of sufficient duration to allow for team-building and integration of research efforts.

• Provide funding mechanisms that facilitate universities working together (including those from different countries) to address societal problems each would be challenged to address alone.

• Develop mechanisms for budgetary flexibility in long-term, multi-institutional grants.

• Acknowledge, for projects that require more than a single principal investigator (PI), the equal leadership status of multiple PIs when "co-PI" is ambiguous.

• Remove administrative barriers to, and explicitly encourage, partnerships between universities, industry, and federal laboratories to facilitate IDR.

**F-2: Funding organizations, including interagency cooperative activities, should provide mechanisms that link interdisciplinary research and education and should provide opportunities for broadening training for researchers and faculty members.**

They can

• Require institutions that receive IDR funding to demonstrate support for interdisciplinary educational activities, such as team teaching.

• Provide, to the extent allowed by the funding organization's mission and guidelines, special grants to support interdisciplinary teaching.

• Designate funds for IDR meetings that encourage interaction between researchers in different disciplines so they can learn about the research in other fields and network with other researchers with whom they might collaborate.

• Support sabbaticals and leaves of absence for studies that focus on interdisciplinary scholarship.

- Ensure that their staff is knowledgeable about interdisciplinarity.

**F-3: Funding organizations should regularly evaluate, and if necessary redesign, their proposal and review criteria to make them appropriate for interdisciplinary activities.**

For example, funding organizations can

- Develop criteria to ensure that proposals are truly interdisciplinary and not merely adding disciplinary participants.
- Encourage IDR proposals that fall within the compass of the organizations' overall missions even if they cross internal organizational boundaries or do not fit specific (review) divisions.
- If they are organized along disciplinary lines, develop policies and practices for funding research that may have a major impact on research in other disciplines, for example, by awarding a mathematics section grant to a mathematician to work on a life-sciences project.

**F-4: Congress should continue to encourage federal research agencies to be sensitive to maintaining a proper balance between the goal of stimulating interdisciplinary research and the need to maintain robust disciplinary research.**

## Professional Societies

**PS-1: Professional societies should seek opportunities to facilitate IDR at regular society meetings and through their publications and special initiatives.**

For example, they can

- Include IDR presentations and sessions at regular society meetings by
  — Choosing IDR topics for some of the seminars, workshops, and symposia.
  — Promoting networking and other opportunities to identify potential partners for interdisciplinary collaboration.
  — Cohosting symposia with other societies.
  — Holding workshops on communication skills, leadership, consensus-building, and other skills useful in leading and being part of IDR teams.
- Establish special awards that recognize interdisciplinary researchers.
- Sponsor lectureships that bring recognition of the value of interdisciplinary experience.

•   Prepare glossaries, primers, tutorials, and other materials to assist scientists in other fields who wish to learn new disciplines.

•   Create sections, divisions, or boards that represent interdisciplinary aspects of their fields.

## Journal Editors

**J-1: Journal editors should actively encourage the publication of IDR research results through various mechanisms, such as editorial-board membership and establishment of special IDR issues or sections.**

In particular, journal editors can

•   Increase the exposure of IDR by devoting special issues or sections to specific IDR directions in a field and accepting more research papers that introduce new IDR areas.

•   Add researchers with interdisciplinary experience to editorial boards and review panels and develop specific techniques for evaluating interdisciplinary submissions.

•   Consider whether their publications' guidelines for authorship and submission of manuscripts are appropriate for IDR.

•   Take steps to improve the sharing of knowledge between disciplines by publishing

— Comprehensive review articles on related disciplines.

— Overview articles on fields relevant to published interdisciplinary works.

— A list of the fields covered in interdisciplinary research papers.

— Hyperlinked text in papers directing on-line readers to discipline-specific educational resources.

— Create subscription models based on article title and subject rather than journal title to enhance cross-discipline access.

## Evaluation of IDR

**E-1: IDR programs and projects should be evaluated in such a way that there is an appropriate balance between criteria characteristic of IDR, such as contributions to creation of an emerging field and whether they lead to practical answers to societal questions, and traditional disciplinary criteria.**

For example, organizations that review IDR can measure

•   The degree to which IDR contributes to the creation of an emerging field or discipline; emerging fields have included nanoscience and nanotechnology and cognitive science.

• How well IDR enhances the training of students and the careers of researchers in ways that surpass the results expected from disciplinary research; these might include employment in a broader array of positions, more rapid progress in gaining tenure and other goals, and greater numbers of speaking invitations.

• Whether the research leads to practical answers to societal questions; for example, an IDR effort to reduce hunger should produce some measurable progress toward that goal. The same IDR program might produce additional outcomes of value, including basic research, that were not expected.

• Whether participants demonstrate an expanded research vocabulary and abilities to work in more than one discipline.

• The extent to which IDR activities, institutes, or centers enhance the reputation of the host institutions; reputation can be measured in research funding, external recognition of IDR leadership, awards, and recognition of participants in the research.

• The long-term productivity of a program; not all initiatives will have the same lifetime, and the use of "sunset" provisions should be considered in the planning of IDR centers and programs.

• Adopt multiple measures of research success, as appropriate to the fields being evaluated, such as conference presentations or patents in addition to publication in peer-reviewed journals.

**E-2: Interdisciplinary education and training programs should be evaluated according to criteria specifically relevant to interdisciplinary activities, such as number and mix of general student population participation and knowledge acquisition, in addition to the usual requirements of excellence in content and presentation.**

For example, organizations reviewing interdisciplinary education and training programs can begin with such criteria as the following, to be supplemented with others appropriate to the organizations' missions:

• Are interdisciplinary courses attracting more of the general student population to science and engineering courses?

• Are interdisciplinary courses and programs attracting a new or different mix of students to careers in science and engineering?

• Are interdisciplinary courses effective in instilling scientific and technological literacy and awareness of the roles of science and technology in modern life?

**E-3: Funding organizations should enhance their proposal-review mechanisms so as to ensure appropriate breadth and depth of expertise in the review of proposals for interdisciplinary research, education, and training activities.**

For example, organizations that fund IDR could

• Involve researchers who have experience with and are knowledgeable about interdisciplinarity and ensure representation of the most important disciplinary points of view on panels that review IDR proposals. Evaluate a proposal to its cell-biology research program by using researchers in cell biology and including a substantial number in chemistry, physics, computer science, the social sciences, and the humanities as appropriate; this practice would help to ensure disciplinary breadth and reduce bias.

• Review a proposed interdisciplinary program in climate change by using input not only from experts in climate change and related fields—such as oceanography, meteorology, atmospheric chemistry, and land use—but also experts in the constituent disciplines—such as physics, chemistry, and statistics—and nonresearchers for whom the research is relevant; the contributions of different disciplines might be submitted separately in written form, and these reports would be offered to a full review panel, including both disciplinary and interdisciplinary researchers, for final assessment.

**E-4: Comparative evaluations of research institutions, such as the National Academies' assessment of doctoral programs and activities that rank university departments, should include the contributions of interdisciplinary activities that involve more than one department (even if it involves double-counting), as well as single-department contributions.**

For example, organizations that evaluate such institutions can

• Survey emerging interdisciplinary fields to identify demographic information (e.g., numbers and characteristics of participants in various interdisciplinary fields, and/or the kinds of activities they engage in).

• Experiment with "matrix evaluation" to capture the activities and accomplishments of interdisciplinary researchers; a matrix approach is one that would consider IDR as an integral part of the disciplines in which the researchers are "embedded" and make visible the cross-departmental efforts of the researchers who make up the interdisciplinary teams.

• Include as evaluation criteria the comentoring of doctoral students, the contributions of individuals to multiple departments, and publication criteria. The publication criteria might include the nature of the journal audiences for whom the work is published; citation analysis that reveals a broad, interdisciplinary interest in the work being cited; "double counting" of publications, by which credit for a given paper is awarded to all coauthors; multiple-authorship patterns that would reveal the disciplinary backgrounds of coauthors; and others that are still being developed.

- Include the facilitation of interdisciplinarity as part of the accreditation process.

## Academic Institutional Structure

**U-1: Institutions should explore alternative administrative structures and business models that facilitate IDR across traditional organizational structures.**

For example, institutions can

- Experiment with alternative administrative structures, such as the matrix model, in which people move freely among disciplinary departments that are bridged and linked by interdisciplinary centers, offices, programs, and curricula or, alternatively, create institutions "without walls" that have no disciplinary departments and are organized around problems rather than disciplines.
- Create numerous interdisciplinary courses for mentors, provide graduate students with multiple mentors, and offer faculty numerous opportunities for continuing education.
- Oversee interdisciplinary programs at the university level rather than that of a single college.
- Review programs periodically with the option of terminating those no longer of high priority so that there is flexibility to respond to emerging opportunities.

**U-2: Allocations of resources from high-level administration to interdisciplinary units, to further their formation and continued operation, should be considered in addition to resource allocations of discipline-driven departments and colleges. Such allocations should be driven by the inherent intellectual values of the research and by the promise of IDR in addressing urgent societal problems.**

For example, institutions can

- Put in place policies that allow the return of some indirect cost revenues to research units such that interdisciplinary centers and programs with external funding are not disadvantaged.
- Provide support for graduate students who choose to study interdisciplinary fields with mentoring by more than a single faculty member.

• Provide support for generative technologies (for example, shared whiteboard software) that allow the sharing of information and ideas (for example, drawings, slides, or equations) virtually.

• Invest federal funds in activities that lead to the design and implementation of research activities that take full advantage of a distributed information technology infrastructure to coordinate research across institutional lines.

**U-3: Recruitment practices, from recruitment of graduate students to hiring of faculty, should be revised to include recruitment across department and college lines.**

For example, institutions can

• Admit graduate students into broad fields (for example, biological sciences as opposed to microbiology; engineering as opposed to mechanical engineering) with no requirement to specialize until the end of the first or second year.

• Increase the number of joint faculty appointments and PhD programs from a few to many.

• Recruit faculty for positions in both programs and departments so they can teach both within the special sphere of a program and in foundation courses in traditional areas.

**U-4: The traditional practices and norms in hiring of faculty and in making tenure decisions should be revised to take into account more fully the values inherent in IDR activities.**

For example, institutions can

• Provide robust mechanisms for allocating faculty positions to areas of IDR.

• Provide cross-departmental mechanisms for tenure and promotion review.

• Monitor a tenure-track faculty member's progress toward tenure with both mentors from the faculty member's program and senior faculty in traditional fields of special interest to that faculty member.

**U-5: Continuing social science, humanities, and information-science-based studies of the complex social and intellectual processes that make for successful IDR are needed to deepen the understanding of these processes and to enhance the prospects for the creation and management of successful programs in specific fields and local institutions.**

# Appendixes

# A

# Biographical Information on Members and Staff of Committee on Facilitating Interdisciplinary Research

**NANCY C. ANDREASEN** (Co-chair) is the Director of The MIND Institute in Albuquerque, N.M.; Adjunct Professor of Psychiatry, Neuroscience, and Neurology at the University of New Mexico; and Andrew H Woods Chair of Psychiatry at the University of Iowa in Iowa City. After obtaining a Ph.D. in English literature, Dr. Andreasen became an Assistant Professor of English before turning to medicine. She obtained her MD in 1970 from the University of Iowa and completed her residency training there. Her research interests include multiple aspects of neuroscience and psychiatry. She has conducted studies of creativity, mood disorders, and schizophrenia. She currently applies multimodality neuroimaging tools, including structural Magnetic Resonance (sMR), functional Magnetic Resonance (fMR), and positron emission tomography (PET) to the study of normal brain development and degeneration and to illnesses such as schizophrenia. She leads an interdisciplinary team that includes cognitive neuroscientists, computer scientists, electrical and biomedical engineers, physicists, and physicians. Dr. Andreasen has won numerous honors and awards, the highest of which is the President's National Medal of Science, presented to her in 2000 for her work in biological sciences. She received the Interbrew-Baillet Latour Heath Prize from the Belgian National Foundation for Scientific Research in 2003 for her work in neuroimaging and schizophrenia. She has received the Rhoda and Bernard Sarnat Award from the Institute of Medicine. She also won the Lieber prize for her research in schizophrenia. Other prizes and awards include Woodrow Wilson and Fulbright Fellowships; Honorary Fellow of the RCSP (Canada); Member of the Institute of Medicine; Research Scientist Award from NIMH; Menninger Award for Psychiatric Research; American Psychiatric Association Prize for Research; the Adolph Meyer Award; the Sigmund Freud Award, and the Distinguished Service and Stanley Dean Awards from the American College of Psychiatrists. She is the author of numerous scientific and scholarly articles and fourteen books, ranging from *John Donne: Conservative Revolutionary* (Princeton, 1976) to *Brave New Brain: Conquering Mental Illness in the Era of the Genome* (Oxford, 2001). She has also authored two widely used textbooks on psychiatry and is Editor in Chief of the *American Journal of Psychiatry.*

**THEODORE L. BROWN** (Co-chair) is founding director emeritus and professor emeritus of chemistry at the University of Illinois—Urbana Champaign (UIUC). Dr. Brown received his Ph.D. from Michigan State University in 1956. He has been a faculty member in the UIUC Department of Chemistry since 1956 (he assumed emeritus status in January 1994). During 1980-1986, he served as vice chancellor for research and dean of the Graduate College. He was the first director of the Beckman Institute in 1987-1993. He served as interim vice-chancellor for academic affairs during 1993. He is an emeritus member of the Beckman Institute Advanced Chemical Systems Group. He participated in the National Academies Government-University-Industry Research Roundtable from 1989 to1994. Dr. Brown's fields of research interests were inorganic chemistry and organometallic chemistry, with an emphasis on the kinetics and mechanisms of reactions. His current interests are in the cognitive, philosophic, and social aspects of the scientific enterprise. His recent book *Making Truth: Metaphor in Science* (*http://www.press.uillinois.edu/s03/brown.html*) explores the metaphoric foundations of science. He is a fellow of AAAS (1987) and of the American Academy of Arts and Sciences (1994), received the American Chemical Society Award for Distinguished Service in the Advancement of Inorganic Chemistry (1993), and was a Guggenheim fellow (1979-1980).

**JENNIFER CHAYES** is an expert in the emerging field at the interface of mathematics, physics, and theoretical computer science. She is cofounder and comanager of the Theory Group at Microsoft Research. Dr. Chayes is also an affiliate professor of mathematics and physics at the University of Washington and was for many years a professor of mathematics at the University of California, Los Angeles (UCLA). She is the recipient of a National Science Foundation postdoctoral fellowship, a Sloan fellowship, and the UCLA Distinguished Teaching Award. Dr. Chayes serves on numerous boards, advisory committees, and editorial boards, including the scientific boards of Banff International Research Station and the Fields Institute, the Advisory Boards of the Center for Discrete Mathematics and Computer Science, and the National Academy of Sciences Office for the Public Understanding of Science. She is the chair of the mathematics section of the American Association for the Advancement of Science and is a past vice-president of the American Mathematical Society. Dr. Chayes did her doctoral work in mathematical physics at Princeton and held postdoctoral positions in mathematics and physics at Harvard and Cornell. She has twice been a member of the Institute for Advanced Study in Princeton.

**STANLEY COHEN** is professor and former chair of genetics and professor of medicine at Stanford University. In 1973, he and Herbert Boyer, of the University of California, San Francisco, invented the technique of DNA cloning, which allowed genes to be transplanted between different species. Their discovery signaled the birth of genetic engineering. He received his B.A. magna cum laude in biological sciences from Rutgers University and his M.D. from the University of Pennsylvania. Dr. Cohen's numerous honors and awards include the National Medal of Science, the National Medal of Technology, and the Albert Lasker Basic Medical Research Award.

**JONATHAN R. COLE,** John Mitchell Mason Professor of the University and Provost and Dean of Faculties, Emeritus, received a B.A. in American history from Columbia in 1964 and a Ph.D. with honors in sociology from Columbia in 1969. He has been teaching at Columbia since 1966. He served as director of the Center for the Social Sciences from 1979 to 1987, when he became vice president for arts and sciences, a post he held until July 1989, when he became provost. Among his many awards and honors, he has received a John Simon Guggenheim Fellowship, has been a fellow at the Center for Advanced Study of the Behavioral Sciences, and is a member of the American Academy of Arts and Sciences. Dr. Cole has published extensively on historical and social aspects of science; has been a leading international contributor to the understanding of the opportunities, challenges, and obstacles facing women in the scientific community; has led a National Academy of Sciences evaluation of the peer-review system in science; and has published works recently on health risks and on dilemmas facing American research universities.

**ROBERT CONN** is managing director of Enterprise Partners Venture Capital. He is helping to lead the $350 million Enterprise Partners VI fund, which is targeted to provide early-stage investments in semiconductors, computing, networking, technology-based life-sciences and drug discovery, and enterprise software. He was previously the dean of the University of

California, San Diego (UCSD) Jacobs School of Engineering from 1993 to 2002, and before that served as a professor of engineering and applied science at the University of California, Los Angeles and the University of Wisconsin-Madison. During his tenure as dean, the Jacobs School rose to become ranked among the top 10 public engineering schools in the country. Dr. Conn led efforts to establish major enterprises in key technical areas including the Center for Wireless Communications in 1995 and the California Institute for Telecommunications and Information Technology in 2000. The latter involved a significant partnership between the state of California, the University of California, and industry, with the state contributing $100 million and industry $140 million. He also helped UCSD to win the highly competitive National Partnership for Advanced Computational Infrastructure and the Distributed Terascale Facility at the San Diego Supercomputer Center. Most recently, he established the Jacobs School's William J. von Liebig Center for Entrepreneurism and Technology Advancement, enabled by a $10 million gift from the William J. von Liebig Foundation. Dr. Conn has been a leader in plasma physics, materials research, and fusion-energy development. He has served on many National Academy of Engineering and Department of Energy (DOE) committees and was chair of DOE's primary fusion-energy advisory committee from 1992 through 1996. In the late 1980s, Dr. Conn cofounded a startup company, Plasma and Materials Technologies (PMT), to develop and market semiconductor etching and deposition equipment. Dr. Conn served as chairman of the Board and senior technologist in 1986-1994 and stepped down from affiliation with the company after joining UCSD as dean of engineering. PMT merged in 1997 into what is now Trikon Technologies, headquartered in the UK.

**MILDRED DRESSELHAUS** is Institute Professor of Electrical Engineering and Physics at the Massachusetts Institute of Technology. She has been active in the study of a wide array of problems in the physics of solids. Her recent interests have been nanoscience, carbon nanotubes, nanowires, and low-dimensional thermoelectricity. Dr. Dresselhaus is a member of the American Philosophical Society (APS) and a fellow of the American Academy of Arts and Sciences, the American Physical Society (APS), the The Institute of Electrical and Electronics Engineers (IEEE), the Materials Research Society, the Society of Women Engineers, and the American Association for the Advancement of Science (AAAS). She has served as president of APS, treasurer of the National Academy of Sciences, president of the AAAS, and as a member of numerous advisory committees and councils. She is now chair of the Board of the American Institute of Physics. Dr. Dresselhaus has received numerous awards, including the National Medal of Science and 18 honorary doctorates. She is the coauthor of four books on carbon science.

**GERALD HOLTON** is Mallinckrodt Research Professor of Physics and Research Professor of History of Science at Harvard University. He received his Ph.D. from Harvard in 1948, and his chief interests are the history and philosophy of science, the physics of matter at high pressure, and the study of career paths of young scientists. His books include *Thematic Origins of Scientific Thought* (1973; rev. ed., 1988); *Science and Anti-Science* (1993); *The Advancement of Science, and its Burdens* (1998); *Scientific Imagination* (1998); and *Einstein, History, and Other Passions* (2000). In addition to teaching at Harvard University since 1947, Dr. Holton was a visiting professor at MIT from 1976 to 1994 as a founding faculty member of the Program on Science, Technology and Society. He has been a visiting professor at Leningrad University, the University of Rome, the Centre National de la Recherché Scientifique (Paris), and Imperial College (London) and a lecturer in China for the Chinese Academy of Social Science. He has been an officer of numerous professional organizations, including president of the History of Science Society (1983-1984), vice president of the Académie Internationale d'Histoire des Sciences (1981-1988), and founding chairman of the American Institute of Physics Committee for the Center for History of Physics. Dr. Holton is a fellow of the American Physical Society, the American Philosophical Society, the American Academy of Arts and Sciences, and the American Association for the Advancement of Science. His awards include the Sarton Medal (1989) and the Joseph H. Hazen Prize (1998) of the History of Science Society, the J.D. Bernal Prize of the Society for Social Studies of Science (1989), the Andrew Gemant Award of the American Institute of Physics (1989), the Joseph Priestley Award of Dickinson College (1994), the Oersted Medal of the American Association of Physics Teachers (1980), and selection as a Jefferson Lecturer by the National Endowment for the Humanities (1981).

**THOMAS KALIL** is the special assistant to the chancellor for science and technology at the University of California, Berkeley and an adjunct fellow at the New America Foundation. At Berkeley, he is helping faculty members to develop research and education initiatives that respond to national priorities and that build strong partnerships with government agencies, the private sector, and community-based organizations. He previously coordinated technology policy for the National Economic Council during the Clinton administration and has served as a consultant to the Digital Promise project. He was a trade specialist at the Washington offices of Dewey Ballantine, where he represented the Semiconductor Industry Association on U.S.-Japan trade issues and technology policy. He received a B.A. in political science and international economics from the University of Wisconsin-Madison and completed graduate work at the Fletcher School of Law and Diplomacy. He is the author of articles on nuclear strategy, U.S.-

Japan trade negotiations, U.S.-Japan cooperation in science and technology, the National Information Infrastructure, distributed learning, and electronic commerce. He is a member of the Council on Foreign Relations, the Association for Computing Machinery, the Internet Society, and the Institute for Electrical and Electronics Engineers.

**ROBERT W. KATES** is a geographer and independent scholar in Trenton, Maine, and university professor (emeritus) at Brown University. His current research focuses on long-term trends in environment, development, and population. He is co-convenor of the international Initiative for Science and Technology for Sustainability, an executive editor of *Environment* magazine, and visiting scholar at the Belfer Center for Science and International Affairs, Kennedy School of Government, Harvard University. Dr. Kates developed and directed three academic interdisciplinary centers: in resource assessment at the University of Dar Es Salaam; on technology, environment, and development at Clark University; and on World Hunger at Brown University. He is a recipient of the 1991 National Medal of Science and the MacArthur Prize Fellowship (1981–85) and is a member of the American Academy of Arts and Sciences and of the Academia Europaea. Dr. Kates received an M.A. and a Ph.D. in geography from the University of Chicago and an honorary D.Sc. from Clark University.

**TIMOTHY L. KILLEEN** was born in Cardiff, Wales. He received a B.Sc. in physics and a Ph.D. in atomic and molecular physics from University College, London. He is director of the National Center for Atmospheric Research (NCAR) in Boulder, Colorado, and a senior scientist at the NCAR High Altitude Observatory, where he leads an experimental and theoretical program in upper atmospheric research. Before joining NCAR, Dr. Killeen was professor of atmospheric and space sciences at the University of Michigan. During his tenure at Michigan, he was also director of the Space Physics Research Laboratory and associate vice president for research. He has taught many undergraduate and graduate courses, including an innovative introductory course sequence for nonscience majors dealing with the physical and human impacts of global change. He has been honored with the Excellence in Teaching and Excellence in Research awards from the College of Engineering at the University of Michigan and with two National Aeronautics and Space Administration (NASA) achievement awards. His research interests include the experimental and theoretical study of the earth's upper atmosphere. He is a principal investigator and instrument developer for a spaceborne Doppler interferometer on the NASA TIMED spacecraft. He is co-principal investigator for a new National Science Foundation (NSF) science and technology center devoted to numerical modeling of space weather. Dr. Killeen has served as president of the Space Physics

Section of the American Geophysical Union and on various NASA and NSF committees. He is editor-in-chief of the *Journal of Atmospheric and Solar-Terrestrial Physics*.

**MARIO MOLINA** has been involved in developing our understanding of the chemistry of the stratospheric ozone layer and its susceptibility to human-made perturbations. In 1974, Dr. Molina and F. S. Rowland reported in *Nature* on their research on the threat to the ozone layer from chlorofluorocarbon gases that were being used as propellants in spray cans, as refrigerants, as solvents, and so on. More recently, he has been involved with the chemistry of air pollution of the lower atmosphere. He is also pursuing interdisciplinary work on tropospheric pollution, working with colleagues in many other disciplines on the problem of rapidly growing cities with severe air pollution. Dr. Molina was born in Mexico City, Mexico. He holds a degree in chemical engineering (1965) from the Universidad Nacional Autonoma de Mexico; a postgraduate degree (1967) from the University of Freiburg, Germany, and a Ph.D. in physical chemistry (1972) from the University of California, Berkeley. He went to the Massachusetts Institute of Technology (MIT) in 1989 with a joint appointment in the Department of Earth, Atmospheric and Planetary Sciences and the Department of Chemistry and was named MIT institute professor in 1997. Before joining MIT, he held teaching and research positions at the Universidad Nacional Autonoma de Mexico; the University of California, Irvine; and the Jet Propulsion Laboratory at the California Institute of Technology. Dr. Molina is a member of the United States National Academy of Sciences and the Institute of Medicine, and the Pontifical Academy of Sciences. He has served on the U.S. president's Committee of Advisers in Science and Technology, the secretary of energy advisory board, the National Research Council Board on Environmental Studies and Toxicology, and the boards of the U.S.-Mexico Foundation of Science and other nonprofit environmental organizations. He has received several awards for his scientific work, including the 1995 Nobel Prize in chemistry, which he shared with F. S. Rowland and P. Crutzen for their work in atmospheric chemistry.

**PATRICK SUPPES** is the Lucie Stern Professor of Philosophy Emeritus at Stanford University and since 1992 has been the director and faculty adviser of Stanford's Education Program for Gifted Youth. He was director of Stanford's Institute for Mathematical Studies in the Social Sciences (1959-1992). He is also professor emeritus by courtesy in Stanford's Department of Statistics, Department of Psychology, and School of Education. Dr. Suppes is a fellow of the American Association for the Advancement of Science (1962), the American Psychological Association (APA) (1964), and the American Academy of Arts and Sciences (1968) and is a member of the National Academy of Education (NAE) (1965), and a member of the American Philosophical Society (1991). Among his awards are the APA Distinguished Scientific Contribution Award, the Columbia University Teachers College Medal for Distinguished Service (1978) and the National Medal of Science (1990). He is a past president of the Pacific Division of American Philosophical Association (1972-1973), the American Educational Research Association (1973-1974), NAE (1973-1977), and the International Union of History and Philosophy of Science (1976, 1978). Dr. Suppes received his bachelor's degree from the University of Chicago and his doctorate from Columbia University. He has published widely in philosophy, the social sciences, and education.

**JAN H. VAN BEMMEL** is professor of medical informatics, first at Free University Amsterdam, 1973-1987, thereafter at Erasmus University Rotterdam, the Netherlands, 1987. He was rector magnificus (vice chancellor) of Erasmus University, Rotterdam, 2000-2003. He received his M.Sc. in physics and mathematics from Technical University Delft in 1963, and his Ph.D. in physics and mathematics from Nijmegen University in 1969. He has been editor-in-chief of *Methods of Information in Medicine*, 1986-2001, of the *IMIA Yearbooks of Medical Informatics*, 1992-2001, and of the *Handbook of Medical Informatics*, 1995-97. He was President of the International Medical Informatics Association, 1998-2001. He became a member of Royal Netherlands Academy of Arts and Sciences (KNAW), 1987, member of Dutch Health Council, 1987, and foreign associate member of Institute of Medicine of National Academy of Sciences, 1991. He was chairman of the International Committee of KNAW for the assessment of all biomedical and health sciences research in the Netherlands, 1993-1998, and chairman of the KNAW Committee for the future assessment of all university research in the Netherlands, 1999-2001.

**TANDY WARNOW** is Professor of Computer Sciences at the University of Texas at Austin, where she is a member of five graduate groups (Computer Sciences, Mathematics, Computational and Applied Mathematics, Molecular Biology, and Ecology, Evolution, and Behavior). Her research combines mathematics, computer science, and statistics to develop improved models and algorithms for reconstructing complex and large-scale evolutionary histories in both biology and historical linguistics. She is on the board of directors of the International Society for Computational Biology and previously was the Co-Director of the Center for Computational Biology and Bioinformatics at the University of Texas at Austin. She received the National Science Foundation Young Investigator Award in 1994, and the David and Lucile Packard Foundation Award in Science and Engineering in 1996.

**ROBERT M. WHITE** is university professor of electrical and computer engineering and director of the Data Storage Systems Center at Carnegie Mellon University (CMU). He received a B.S. in 1960 from the Massachusetts Institute of Technology and a Ph.D. in 1964 from Stanford University. In addition to an active program of research in data-storage systems, Dr. White has longstanding interests in technology policy. His policy interests are focused on federal science and technology policy. He is exploring the effects of various government policies on technology innovation, whereby new technology appears in a competitive product or process. Examples of issues include the effects of federal funding and the management of intellectual property. Before joining CMU, he served during the first Bush administration as the first undersecretary of commerce for technology. Earlier, he was vice president of Microelectronics and Computer Technology Corporation (MCC). He was a manager and a principal scientist at Xerox PARC and then moved on to serve as vice president of Control Data Corporation before his position at MCC. Dr. White's professional memberships include the American Physical Society, the Institute of Electrical and Electronics Engineers, and the American Association for the Advancement of Science. He serves on the boards of directors of several companies, including ST-Microelectronics and Silicon Graphics.

**MARY LOU ZOBACK** is a senior research scientist with the U.S. Geological Survey (USGS) Earthquake Hazards Team in Menlo Park, California. She received a Ph.D. in geophysics from Stanford University in 1978 and joined the USGS earthquake-studies staff permanently in 1979 after a year National Research Council postdoctoral fellowship at USGS. From 1986 to 1992, she led the World Stress Map project, a task group of the International Lithosphere Program that involved 40 scientists in 30 countries in an effort to compile and interpret geological and geophysical data on the present-day tectonic stress field. Dr. Zoback has served on a National Research Council panel to evaluate the proposed Yucca Mountain site for long-term disposal of radioactive waste, on a steering committee for the National Aeronautics and Space Administration (NASA) Solid Earth Sciences program to define 20- to 25-year goals for that program, and on a USGS team to define a 10-year science strategy for the Geologic Division of USGS. She is a past president of the Geological Society of America and served as president of the Tectonophysics Section of the American Geophysical Union (AGU) and as a member of the AGU Council. Her honors include the AGU Macelwane Award (1987), a USGS Gilbert Fellowship Award (1990-1991) for a one-year sabbatical in Karlsruhe, Germany, and the Meritorious Service Award from the U.S. Department of the Interior (2002).

## Professional Staff

**DEBORAH D. STINE** (Study Director) is associate director of the Committee on Science, Engineering, and Public Policy (COSEPUP) and director of the Office of Special Projects. She has worked on various projects at the National Academies since 1989. She received a National Research Council group award for her first study for COSEPUP, on policy implications of greenhouse warming; a Commission on Life Sciences staff citation for her work in risk assessment and management; and two awards from the Policy and Global Affairs Division for her efforts in dissemination of National Academies' reports. Other studies have addressed human reproductive cloning, setting priorities for NSF's large research facilities, science and technology presidential appointments, science and technology centers, international benchmarking of U.S. research fields, graduate and postdoctoral education, responsible conduct of research, careers in science and engineering, and many environmental topics. She holds a bachelor's degree in mechanical and environmental engineering from the University of California, Irvine; a master's degree in business administration; and a Ph.D. in public administration, specializing in policy analysis, from the American University. Before coming to the National Academies, Dr. Stine was a mathematician for the U.S. Air Force, an air-pollution engineer for the state of Texas, and an air-issues manager for the Chemical Manufacturers Association.

**LAUREL HAAK** (Program Officer) is a program officer for the Committee on Science, Engineering, and Public Policy (COSEPUP). She received a B.S. and an M.S. in biology from Stanford University. She was the recipient of a predoctoral National Institutes of Health (NIH) National Research Service Award and received a Ph.D. in neuroscience in 1997 from Stanford University Medical School, where her research focused on calcium signaling and circadian rhythms. She was awarded a National Academy of Sciences (NAS) Research associateship to work at NIH on intracellular calcium dynamics in oligodendrocytes. After working at NIH, she joined the staff at the American Association for the Advancement of Science and was editor of Science's Next Wave Postdoc Network. While a postdoctoral scholar, she was editor of the *Women in Neuroscience* newsletter, and she is now president of this organization. She has served on the Society for Neuroscience Committee for the Development of Women's Careers in Neuroscience and the Biophysics Society Early Careers Committee, and she was an adviser and mentor for the National Postdoctoral Association.

# B

# Charge to the Committee

T he committee conducting this study will examine the scope of interdisciplinary research and provide findings, conclusions, and recommendations as to how such research can be facilitated by funding
organizations and academic institutions. The committee will recognize in
its deliberations that the organization of research in academic institutions is
driven by teaching and other considerations

Specifically, the committee will address the following tasks:

- Review proposed definitions of interdisciplinary research including
similarities and differences from research characterized as cross-disciplinary, intradisciplinary, and multi-disciplinary and develop measures to determine whether research is interdisciplinary or not.
- Identify and analyze current structural models of interdisciplinary
research.
- Identify and analyze the policies and procedures of Congress, funding organizations, and institutions that encourage or discourage interdisciplinary research.
- Compare and contrast current structural models and policies and
procedures in academic and non-academic settings as well as traditional
and non-traditional academic settings that encourage or discourage interdisciplinary research.
- Identify measures that can be used to evaluate the impact on research, graduate students and postdoctoral scholars, and researchers ex-

pected from their engagement in greater interdisciplinary research and cross-professional opportunities.

- Develop findings and conclusions as to the current state of interdisciplinary research and the factors that encourage (or discourage) it in academic, industry, and federal laboratory settings.
- Provide recommendations to academic institutions and public and private sponsors of research as to how to better stimulate and support interdisciplinary research.

# C

# Convocation Program and Speakers Biographies

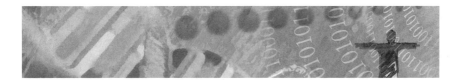

## WELCOME

Welcome to the National Academies' Convocation on Facilitating Interdisciplinary Research.

The purpose of this convocation is to better understand the concerns of funding organizations, university administrators, faculty, researchers, and students regarding interdisciplinary research and to identify effective practices and structural models, policies, and procedures that could help facilitate interdisciplinary research. The convocation consists of four elements:

- A series of panel discussions with federal, private, and international funding organizations, researchers, research center directors, and educators.
- Poster sessions where attendees can share their experiences.
- A public comment session.
- A survey of convocation participants.

The discussions during these activities will help the committee respond to its charge. We encourage you to fully participate in the convocation and we look forward to hearing your ideas.

Thank you again for coming!

## COMMITTEE ON SCIENCE, ENGINEERING, AND PUBLIC POLICY

**MAXINE F. SINGER** (Chair), President Emeritus, Carnegie Institution of Washington

**BRUCE ALBERTS** (Ex-officio), President, The National Academies

**R. JAMES COOK**, R. James Cook Endowed Chair in Wheat Research, Washington State University

**HAILE DEBAS**, Dean, School of Medicine and Vice Chancellor, Medical Affairs, University of California, San Francisco

**GERALD DINNEEN** (Ex-officio), Retired Vice President, Science and Technology, Honeywell, Inc.

**HARVEY FINEBERG** (Ex-officio), President, Institute of Medicine

**MARYE ANNE FOX** (Ex-officio), Chancellor, University of California, San Diego

**ELSA GARMIRE**, Sydney E. Junkins Professor of Engineering, Dartmouth College

**NANCY HOPKINS**, Amgen Professor of Biology, Massachusetts Institute of Technology

**WILLIAM JOYCE** (Ex-officio), Chairman and CEO, Hercules Incorporated

**MARY-CLAIRE KING**, American Cancer Society Professor of Medicine and Genetics, University of Washington

**W. CARL LINEBERGER**, Professor of Chemistry, Joint Institute for Laboratory Astrophysics, University of Colorado

**ANNE PETERSEN**, Senior Vice President for Programs, W.K. Kellogg Foundation, Battle Creek, Michigan

**CECIL PICKETT**, President, Schering-Plough Research Institute

**GERALD RUBIN**, Vice President for Biomedical Research, Howard Hughes Medical Institute

**HUGO SONNENSCHEIN**, Charles L. Hutchinson Distinguished Service Professor, Department of Economics, The University of Chicago

**JOHN D. STOBO**, President, University of Texas Medical Branch of Galveston

**IRVING WEISSMAN**, Karel and Avice Beekhuis Professor of Cancer Biology, Stanford University

**SHEILA WIDNALL**, Abbey Rockefeller Mauze Professor of Aeronautics, Massachusetts Institute of Technology

**WM. A. WULF** (Ex-officio), President, National Academy of Engineering

**MARY LOU ZOBACK**, Senior Research Scientist, Earthquake Hazards Team, U.S. Geological Survey

Staff

RICHARD BISSELL, Executive Director
DEBORAH D. STINE, Associate Director
LAUREL HAAK, Program Officer
MARION RAMSEY, Administrative Associate

## ABOUT THE COMMITTEE ON FACILITATING
## INTERDISCIPLINARY RESEARCH

As part of the National Academies Keck Futures Initiative, the National Academies—under the aegis of the Committee on Science, Engineering, and Public Policy—launched a study to examine how funding organizations and academic institutions can best facilitate interdisciplinary research. The study is funded by the W. M. Keck Foundation.

### Charge to the Committee

The committee conducting this study will examine the scope of interdisciplinary research and provide findings, conclusions, and recommendations as to how such research can be facilitated by funding organizations and academic institutions. The committee will recognize in its deliberations that the organization of research in academic institutions is driven by teaching and other considerations

The Committee on Facilitating Interdisciplinary Research is charged with:

• Reviewing proposed definitions of interdisciplinary research, including similarities and differences from research characterized as cross-disciplinary, interdisciplinary, and multi-disciplinary and develop measures to determine whether research is interdisciplinary or not.

• Identifying and analyzing current structural models of interdisciplinary research.

• Identifying and analyzing the policies and procedures of Congress, funding organizations, and institutions that encourage or discourage interdisciplinary research.

• Comparing and contrasting current structural models and policies and procedures in academic and non-academic settings as well as traditional and non-traditional academic settings that encourage or discourage interdisciplinary research.

• Identifying measures that can be used to evaluate the impact on research, graduate students and postdoctoral scholars, and researchers expected from their engagement in greater interdisciplinary research and cross-professional opportunities.

- Developing findings and conclusions as to the current state of interdisciplinary research and the factors that encourage (or discourage) it in academic, industry, and federal laboratory settings.
- Providing recommendations to academic institutions and public and private sponsors of research as to how to better stimulate and support interdisciplinary research.

**For More Information**

Web site: *nationalacademies.org/interdisciplinary*

E-mail: interdisciplinary@nas.edu

## ABOUT THE W. M. KECK FOUNDATION

Based in Los Angeles, California, the W. M. Keck Foundation was established in 1954 by the late W. M. Keck, founder of the Superior Oil Company. The Foundation's grant making is focused primarily on pioneering efforts in the areas of medical research, science, and engineering. The foundation also maintains a Southern California Grant Program that provides support in the areas of civic and community services with a special emphasis on children.

In May 2003, the National Academies and W. M. Keck Foundation announced a 15-year, $40 million grant from the Keck Foundation to underwrite the "National Academies Keck *Futures Initiative*," a new program designed to realize the untapped potential of interdisciplinary research. The National Academies Keck *Futures Initiative* was created to stimulate new modes of inquiry and break down the conceptual and institutional barriers to interdisciplinary research that could yield significant benefits to science and society.

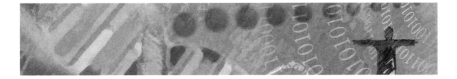

## CONTENTS

## CONVOCATION GUIDELINES

**Questions:** We expect over 300 attendees at the convocation. So that everyone has a chance to ask their questions and provide their comments, we ask that you limit your time at the microphone to one minute. A timing device will be used to ensure we are fair to everyone. When you ask a question or make a comment please state your name and affiliation.

**Survey:** Before you leave we ask you to fill out the survey enclosed in this program and drop it in the box located at the front registration desk. Information from this survey will be used only in aggregate form as part of the committee's data collection efforts.

**Lunch:** Box lunches will be available in the Great Hall directly outside the auditorium. Please take your lunch to one of the following meeting rooms to enjoy. See map below.
Floor 1: 150, 180, Board Room, and Lecture Room
Floor 2: 250 and 280
Committee members and speakers are invited to take meals in the Members' Room located on the first floor.

**Cell phones:** Please either turn off cell phones or place on "vibrate" mode. Messages can be left at (202) 334-1613.

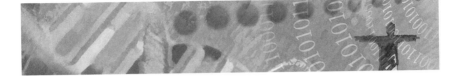

## The National Academies
Committee on Science, Engineering, and Public Policy

## Committee on Facilitating Interdisciplinary Research

## CONVOCATION ON FACILITATING INTERDISCIPLINARY RESEARCH
January 29-30, 2004

National Academy of Sciences Building
2101 Constitution Avenue, NW
Washington, D.C.

### AGENDA

**Thursday, January 29, 2004**

9 AM      Opening Remarks
Nancy Andreasen, Co-Chair, Cmte on Facilitating
Interdisciplinary Research

9:15      Federal Research Funding Agency Perspectives on
Facilitating Interdisciplinary Research
*Moderator:* Mary Lou Zoback, Member, Cmte on
Facilitating Interdisciplinary Research
- Rita Colwell, Director, National Science Foundation
- Ray Orbach, Director, Office of Science, Department of Energy
- William Berry, Director, Basic Research, ODUSD, Department of Defense
- Lawrence Tabak, Director, National Institute of Dental and Craniofacial Research, National Institute of Health
- Cliff Gabriel, Deputy Associate Director, Science Division, White House Office of Science and Technology Policy

10:45      Break

11:00      Private and International Foundation Perspectives on
           Facilitating Interdisciplinary Research
           *Moderator*: **Jonathan Cole**, Member, Cmte on Facilitating
           Interdisciplinary Research
           • **Maria Pellegrini**, Program Director for Science,
             Engineering, and Liberal Arts, W. M. Keck Foundation
           • **Robert Granger**, President, William T. Grant Foundation
           • **Laurie Garduque**, Program Director for Research, John D.
             and Catherine T. MacArthur Foundation
           • **Barry Gold**, Program Officer, Conservation and Science,
             The David and Lucile Packard Foundation
           • **Carmen Charette**, Senior Vice President, Canada
             Foundation for Innovation
           • **Anthony Armstrong**, Director, Indiana 21st Century
             Research & Technology Fund

12:30 PM   Lunch

1:30       Interdisciplinary Researchers' Perspectives on Facilitating
           Interdisciplinary Research
           *Moderator:* **Stan Cohen**, Member, Cmte on Facilitating
           Interdisciplinary Research
           • **F. Sherwood Rowland**, Bren Research Professor,
             Chemistry and Earth System Science, University of
             California at Irvine
           • **Joel Cohen**, Abby Rockefeller Mauzé Professor,
             Laboratory of Populations, Rockefeller University and
             Columbia University
           • **Lee Magid**, Professor, Chemistry, University of
             Tennessee, and Acting Director, Joint Institute for
             Neutron Sciences, UT and Oak Ridge National
             Laboratory
           • **Diana Rhoten**, Program Officer, Social Science Research
             Council
           • **Feniosky Peña-Mora**, Associate Professor of Construction
             Management and Information Technology William E.
             O'Neil Faculty Scholar, Civil and Environmental
             Engineering Department, University of Illinois Urbana-
             Champaign
           • **Victoria Interrante**, Assistant Professor, Computer Science
             and Engineering, University of Minnesota

3:00        Break

3:15        **Research Center Directors' Perspectives on Facilitating Interdisciplinary Research**
            *Moderator:* **Mario Molina**, Member, Cmte on Facilitating Interdisciplinary Research
            • **Harvey Cohen**, Professor, Pediatrics, Stanford School of Medicine, and Chair, The Interdisciplinary Initiatives Committee, Bio-X, Stanford University
            • **Catherine Ross**, Director, Center for Quality Growth, Georgia Tech
            • **Pierre Wiltzius**, Director, Beckman Institute for Advanced Science and Technology, and Professor, Materials Science and Engineering Department and Physics Department University of Illinois at Urbana-Champaign
            • **Uma Chowdhry**, Vice President, Central Research and Development, DuPont
            • **Jeffrey Wadsworth**, Director, Oak Ridge National Laboratory
            • **Ruzena Bajcsy**, Director, Center for Information Technology Research in the Interest of Society, University of California, Berkeley

4:45        Break

5-6:00 PM   Plenary Discussion
            *Moderator:* **Nancy Andreasen**, Co-Chair, Cmte on Facilitating Interdisciplinary Research
            *Discussant:* **Julie Thompson Klein**, Professor of Humanities, Wayne State University

6-7:00 PM   Poster Session

**Friday, January 30, 2004**

9:00 AM     Welcome
            **Theodore Brown**, Co-Chair, Cmte on Facilitating Interdisciplinary Research

            **Perspectives on Education and Training: Creating a New Generation of Interdisciplinary Researchers**
            *Moderator:* **Jennifer Chayes**, Member, Cmte on Facilitating Interdisciplinary Research

- **Marye Anne Carroll**, Professor, Atmospheric, Oceanic, and Space Sciences; Professor, Chemistry; Director, Program for Research on Oxidants: Photochemistry, Emissions, and Transport (PROPHET); Director, Biosphere—Atmosphere Research and Training (BART), University of Michigan
- **Edward Miles**, Professor of Marine Studies and Public Affairs, University of Washington
- **Alice Gottlieb**, Professor of Medicine and Director of the Clinical Research Center, Robert Wood Johnson Medical School, University of Medicine and Dentistry New Jersey
- **James Collins**, Ullman Professor of Biology, Arizona State University
- **Julio de Paula**, Professor of Chemistry, Haverford College

10:45     Break

11:00     Plenary Discussion
          *Moderator:* **Theodore Brown**, Co-Chair, Cmte on
          Facilitating Interdisciplinary Research

11:45     Closing Comments

12:00     Adjourn

*Copies of the PowerPoint presentations will be available
shortly after the Convocation at
http://www.nationalacademies.org/interdisciplinary*

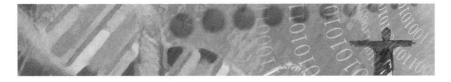

## SPEAKERS BIOGRAPHICAL INFORMATION

**ANTHONY ARMSTRONG** is the Director of the Indiana 21st Century Research and Technology Fund. Prior to joining the Fund, Dr. Armstrong served in the Office of Technology Transfer with Indiana University's Advanced Research and Technology Institute (ARTI). Dr. Armstrong's focus was on the commercialization of innovations from the IU School of Medicine, and with corporate relations on behalf of IU. He was Director of Research with the IU School of Business Johnson Center for Entrepreneurship and Innovation prior to joining ARTI. Dr. Armstrong earned business and law degrees from Indiana University.

**RUZENA BAJCSY** was appointed Director CITRIS (Center for Information Technology Research in the Interest of Society) at the University of California, Berkeley in 2001, where she is also a faculty member in the EECS Department. Prior to coming to Berkeley, she was Assistant Director of the Computer Information Science and Engineering Directorate (CISE) at NSF from 1998 to 2001. Dr. Bajcsy is a pioneering researcher in machine perception, robotics and artificial intelligence. She is former Director of the University of Pennsylvania's General Robotics Automation Sensing Perception Laboratory, which she founded in 1978. She received her master's and Ph.D. degrees in electrical engineering from Slovak Technical University in 1957 and 1967, respectively. She received a Ph.D. in computer science in 1972 from Stanford University. Dr. Bajcsy holds membership in the National Academy of Engineering, the Neuroscience Institute, and the Institute of Medicine. In 2001 she became a recipient of the ACM A. Newell award. She was named to Discover Magazine's November 2002's list of the 50 most important women in science. In April of 2003 she received the CRA Distinguished Service Award and in May 2003 she was named to PITAC (the President's Information Technology Advisory Committee).

**WILLIAM BERRY** is the Director for Basic Research of the Military Services and Defense Agencies. He provides scientific leadership, management oversight, policy guidance and coordination of the $1.2 billion yearly basic research programs. Dr. Berry began his association with the Department of Defense as a National Research Council Postdoctoral Fellow at the Air Force Aerospace Medical Research Laboratory in 1976. Immediately prior to his current position, Dr. Berry was Associate Deputy Assistant Secretary of the Air Force for Science Technology and Engineering and Director of the Washington Office of the Air Force Research Laboratory. His research publications are in the fields of environmental toxicology and neuroscience. Dr. Berry earned a B.S. in Biology from Lock Haven University, Lock Haven, PA, a M.A.T. in Zoology from Miami University, Oxford, OH, and a Ph.D. in Zoology/Biochemistry from the University of Vermont, Burlington, VT. He is a member of the American Association for the Advancement of Science and Sigma Xi, The Scientific Research Society.

**MARY ANNE CARROLL** is a professor of atmospheric science and chemistry and director of the Program for Research on Oxidants: Photochemistry, Emissions and Transport (PROPHET) at the University of Michigan. She is also Director of the NSF Research Experiences for Undergraduates in Atmospheric Chemistry, Meteorology, and Atmosphere–Forest Exchange and Principal Investigator for the Biosphere–Atmosphere Research and Training (BART) Program, a multi-institutional and multidisciplinary program for doctoral students (NSF IGERT). Dr. Carroll's research efforts include instrument development and field measurements focusing on the impacts of global change on atmospheric oxidant photochemistry and atmosphere–forest exchange. Dr. Carroll was a Research Chemist at the National Oceanic and Atmospheric Administration's Aeronomy Laboratory between 1984 and 1992, following a Postdoctoral Fellowship at the University of Colorado's Cooperative Institute for Research in Environmental Sciences. She also served as Associate Director of NSF's Atmospheric Chemistry Program from 1990 to1992 prior to joining the AOSS and Chemistry faculties at UM. During 1997–2000, Dr. Carroll served as Editor of the *Journal of Geophysical Research—Atmospheres*. Dr. Carroll holds a B.A. in Chemistry from the University of Massachusetts at Boston and a Sc.D. in Atmospheric Chemistry from the Massachusetts Institute of Technology.

**CARMEN CHARETTE** first joined the Canada Foundation for Innovation in July 1997 as vice president, programs. A year later, as the Foundation's scope and influence grew within Canada's science and innovation community, she was appointed to the position of senior ice president, program and operations. Today, Ms. Charette continues to play a significant role in carrying out the CFI's mandate and in keeping the Foundation focused on its increasing responsibility to Canada's research community. Before joining the CFI, Ms. Charette held a variety of Director positions during her 13 years at the Natural Sciences and Engineering Research Council (NSERC). She became the first Chair of the NSERC Operations Committee in 1997, and has continued as a member of the NSERC Senior Management committee to strategic planning. In addition, in 1996, she served as Presidenté de l'Association des administratrices et des administrateurs de recherche universitaire du Québec (ADARUQ). Ms Charette holds a B.S. in Biochemistry and a Bachelor of Business Administration, buth from the University of Ottawa.

**UMA CHOWDHRY** is vice president of Central Research & Development (CR&D) at DuPont, where she began in 1977 as a research scientist. For her contributions to the science of ceramics, Dr. Chowdhry was elected "Fellow" of the American Ceramics Society in 1989. For work ranging from catalysts to superconductors, she was elected to the National Academy of Engineering in 1996. Dr. Chowdhry has served on advisory boards of engineering schools at MIT, University of Pennsylvania, Princeton University and the University of Delaware as well as on the program advisory board and election subcommittee for the National Academy of Engineering. She has served on the National Research Council's study groups that generated assessment reports on various technology topics of national interest. She was recently elected to the board of directors for the Industrial Research Institute, the national Inventors' Hall of Fame and to a Laboratory Operations Board for the Department of Energy for the US Government. Dr. Chowdhry is a member of the National Committee on Women in Science and Engineering sponsored by both the National Academy of Science and the National Academy of Engineering since 1999. Born and raised in Mumbai, India, she came to the United States in 1968 with a B.S. in physics from Indian Institute of Science, Mumbai University, received an M.S. from Caltech in engineering science in 1970 and a Ph.D. in materials science from MIT in 1976.

**HARVEY COHEN** is a professor of pediatrics and chief of staff at Lucile Packard Children's Hospital, and has been named the first holder of the Arline and Pete Harman Professorship for the Chair of the Department of Pediatrics in the School of Medicine. Dr. Cohen received both his M.D. and his Ph.D. (biochemistry) in 1970 from Duke University School of Medicine. His postdoctoral work included a pediatrics residency at Children's Hospital in Boston and a pediatric hematology/oncology fellowship at Children's and the Dana Farber Cancer Institute. He held faculty posts at Harvard Medical School and at the University of Rochester School of Medicine, where he was James P. Wilmot Associate Professor of Pediatric Oncology and Associate Chair for Research and Development in the Department of Pediatrics. He was recruited to Stanford in 1993 as chair of the pediatrics department. His research interests include clinical trials in leukemia, mechanisms of drug resistance, immune killing of bacteria and tumor cells, free radical biochemistry and cell biology. He serves on the national Steering

Committee of the Pediatric Scientist Development Program and chairs the Interdisciplinary Initiative Program Committee for Bio-X, a new venture into scientific research, education and innovation at Stanford.

**JOEL E. COHEN** is Professor of Populations at the Rockefeller University and Columbia University, New York. Cohen's research deals with the demography, ecology, epidemiology and social organization of human and non-human populations and with mathematical concepts useful in these fields. Cohen earned two doctorates, a Ph.D. in Applied Mathematics (1970) and a DrPH in Population Science and Tropical Public Health (1973), from Harvard University. Cohen was elected to the American Academy of Arts and Sciences in 1989 (in evolutionary and population biology and ecology), the American Philosophical Society in 1994 (in the professions, arts, and affairs), and the U.S. National Academy of Sciences in 1997 (in applied mathematical sciences). Cohen serves on the Council of the National Academy of Sciences, the Governing Board of the National Research Council, the worldwide Board of Governors of The Nature Conservancy, and the Council of the American Academy of Arts and Sciences, among other boards. He is also a member of the Council on Foreign Relations, New York, and an Honorary Senior Fellow of the Foreign Policy Association, New York. In March 1999, Cohen was named co-winner of the Tyler Prize for Environmental Achievement, and in April 1998, co-winner of the Fred L. Soper Prize of the Pan American Health Organization, Washington, D.C., for work on Chagas' disease.

**JAMES P. COLLINS** is Virginia M. Ullman Professor of Natural History and the Environment in the School of Life Sciences at Arizona State University. From 1989 to 2002 he was Chairman of the Zoology, then Biology Department. Dr. Collins served as Director of the Population Biology and Physiological Ecology program at the National Science Foundation (NSF) in 1985-86. Dr. Collins's research centers on understanding the origin, maintenance, and reorganization of morphological variation within species. A special focus of the research is emerging wildlife diseases and their relationship to the global decline of amphibians; Collins heads an international team of 26 investigators studying this issue. Dr. Collins received his B.S. from Manhattan College and his M.S. and Ph.D. from The University of Michigan. He joined the faculty at Arizona State University in 1975. Dr. Collins is a Fellow of the American Association for the Advancement of Science. He is currently a member and chair of the Advisory Committee to NSF's Assistant Director for Biological Sciences and a member of the Advisory Committee for Environmental Research and Education reporting to NSF's Assistant Director for Geological Sciences.

**RITA R. COLWELL** is the Director of the National Science Foundation. Since taking office, Dr. Colwell has spearheaded the agency's emphases in K-12 science and mathematics education, graduate science and engineering education/training and the increased participation of women and minorities in science and engineering. In her capacity as NSF Director, she serves as Co-chair of the Committee on Science of the National Science and Technology Council. Before coming to NSF, Dr. Colwell was President of the University of Maryland Biotechnology Institute from 1991 to1998, and she remains Professor of Microbiology and Biotechnology (on leave) at the University Maryland. She was a member of the National Science Board from 1984 to 1990. Dr. Colwell previously served as Chairman of the Board of Governors of the American Academy of Microbiology and also as President of the American Association for the Advancement of Science, the Washington Academy of Sciences, the American Society for Microbiology, the Sigma Xi National Science Honorary Society, and the International Union of Microbiological Societies. Dr. Colwell is a member of the National Academy of Sciences, American Academy of Arts and Sciences, and The American Philosophical Society. Dr. Colwell was born in Beverly, Massachusetts, holds a B.S. in Bacteriology and an M.S. in Genetics, from Purdue University, and a Ph.D. in Oceanography from the University of Washington.

**CLIFFORD GABRIEL** is currently serving as Deputy to the Associate Director for Science in the Office of Science and Technology Policy (OSTP). In this position, he helps shape federal science policy in the physical, life, and social sciences. Dr. Gabriel handles issues for OSTP related to agricultural biotechnology, animal and plant health, animal welfare, food safety, plant genomics, pesticides, Gulf War veterans' illnesses, and dioxin. From 1993 to 1996, Dr. Gabriel was Executive Director of the American Institute of Biological Sciences. As Executive Director, he was responsible for all operations of the Institute including publications, contracts and grants, annual meetings, and public policy. Dr. Gabriel received his Ph.D. in plant pathology from the University of Wisconsin-Madison in 1983.

**LAURIE R. GARDUQUE** is the Director for Research in the MacArthur Foundation's Program on Human and Community Development. Her primary responsibilities focus on activities in mental health, juvenile justice, education, and child and youth development. Dr. Garduque joined the Foundation in 1991 after serving as Director of the National Forum on the Future of Children and Families, a joint project of the National Research Council and the Institute of Medicine. From 1984 to 1987, she was Director in charge of governmental affairs and professional liaison for the American Educational Research Association in Washington, D.C. This position followed the year she spent, from 1983 to 1984, as a Congressional Science Fellow in the U.S. Senate. From 1980 to 1983, Garduque held a faculty position as an Assistant Professor in human development at Pennsylvania State University. She received her bachelor's degree in psychology and her M.A. and Ph.D. in educational psychology from the University of California at Los Angeles.

BARRY GOLD is Program Officer for Conservation and Science at the The David and Lucile Packard Foundation and in this role leads the Foundation's efforts to develop and implement two new strategies. The first is intended to foster the development of the emerging field of sustainability science, while the second will guide scientific activities in support of the Foundation's Oceans and Coasts program. Before joining the Foundation, Dr. Gold led an effort to understand and protect some of the most highly prized scenic and natural resources in the United States while balancing potentially conflicting social and political interests and demands upon the resource. Dr. Gold has dedicated his career to working at the environmental science and policy interface. In this role he has advised senior officials in Congress, federal and state agencies, the White House, non-governmental organizations and civic groups. Dr. Gold holds a D.Sc. from Washington University, an M.A. from George Washington University, an M.S. from the University of Connecticut, and a B.S. from the University of Miami. He is a member of AAAS, the Ecological Society of America, and Sigma Xi.

ALICE GOTTLIEB has spent the majority of her professional career treating and researching immunology and inflammatory diseases and disorders. Her own passion for research, coupled with a desperate need for clinical research into these conditions, prompted her to develop a research fellowship program for promising physicians. She is currently a Professor of Medicine, director of the Clinical Research Center at UMDNJ-Robert Wood Johnson Medical School and holds the W. H. Conzen Chair in Clinical Pharmacology at UMDNJ-Robert Wood Johnson Medical School. Dr. Gottlieb received her medical degree from Cornell University Medical College in 1980, her Ph.D. in Immunology from the Rockefeller University in 1977 and completed her residency at New York Hospital and was certified by the American Board of Dermatology in 1993. She is also board certified in Rheumatology (1984) having trained at the Hospital for Special Surgery and board certified in Internal Medicine (1982) having trained at the New York Hospital.

ROBERT GRANGER is President of the William T. Grant Foundation. Since joining the Foundation in 2000, Dr. Granger has been responsible for leading the Foundation's grantmaking, including refinements that would improve its impact on youth policy and practice. He came to the Foundation from the Manpower Demonstration Research Corporation (MDRC), where he was senior vice president and director of MDRC's education, children, and youth department. Prior to that he was executive vice president at the Bank Street College of Education, and executive director of the Child Development Associate National Credentialing Program. Dr. Granger's research specialties include the study of social programs and policies that affect low-income children, youth, and families. He earned his doctorate in education from the University of Massachusetts, Amherst.

**VICTORIA INTERRANTE** is Assistant Professor in the Department of Computer Science and Engineering at the University of Minnesota. Her research focuses on the application of insights from visual perception, art and illustration to the design of more effective techniques for conveying data through images. Her research involves active collaborations with colleagues across the University from the Department of Aerospace Engineering and Mechanics to the Department of Architecture. Her present projects include: the study of texture's effect on shape perception and the design and synthesis of texture patterns to facilitate accurate shape representation; the study of texture perception and the development of methods for effectively using texture in visualizing multivariate data and representing data uncertainty; the development of algorithms for the effective detection, tracking and visualization of vortical structures in turbulent boundary layer flows; and the study of spatial perception in immersive virtual environments and the use of VR technology in the development of tools to enhance the process of conceptual design in architecture. She received her B.A. in computer science from the University of Massachusetts at Boston in 1984, her M.S. from UCLA in 1986, and her Ph.D. in 1996 from the University of North Carolina at Chapel Hill, where she studied under the joint direction of Dr. Henry Fuchs and Dr. Stephen Pizer. From 1996-1998 she worked as a staff scientist at ICASE, a non-profit research center operated by the Universities Space Research Association at NASA Langley. In 1999 she received a Presidential Early Career Award for Scientists and Engineers, and she was awarded a 2001-2003 McKnight Land-Grant Professorship from the University of Minnesota.

**JULIE THOMPSON KLEIN** is an internationally recognized scholar in the field of interdisciplinary history, theory, and methodology. Dr. Klein arrived at Wayne State in 1970 and has been with what is now the Department of Interdisciplinary Studies in the College of Lifelong Learning since 1976. A past president of the Association for Integrative Studies, she lectures and consults throughout the world for universities developing interdisciplinary programs. Professor Klein currently is a member of the Association of

American Colleges and Universities national task force on Integrative Learning.

**LINDA J. (LEE) MAGID** is a Professor of Chemistry at the University of Tennessee. Her research focuses on physiochemical investigations of micelles and polyelectrolytes in aqueous solutions; techniques used include light scattering, small-angle neutron scattering, neutron spin-echo spectroscopy and NMR spectroscopy. She has served as Vice-President for Research and Graduate Studies at the University of Kentucky and is currently UT's ORNL/SNS Liaison for Science & Technology and the Acting Director of the UT/ORNL Joint Institute for Neutron Sciences. She has a B.S. in chemistry from Rice University and a Ph.D. in chemistry from the University of Tennessee. She is a Fellow of AAAS. Currently she is a member of the NRC Board on Physics and Astronomy and serves as vice-chair of the Solid State Sciences Committee. In addition, she serves on the Board on Assessment of NIST Programs' subpanel on the NIST Center for Neutron Research, and on the U.S. National Committee to the IUPAC. She also served on the Committee on Developing a Federal Materials Strategy.

**EDWARD L. MILES** is the author of many studies on international organizations, international science and technology policy, and marine policy and ocean management. He has served as chairman of the Ocean Policy Committee, National Academy of Sciences/National Research Council (1974-79); member of the Executive Board, Law of the Sea Institute, 1972-81 and 1985-89 and President 1989-93; Chairman of the Legal and Institutional Task Group on the Implications of Disposal of High-Level Radioactive Waste into the Seabed and Advisor to the Executive Committee, Seabed Working Group, Nuclear Energy Agency, OCED, 1981-1987; and Chairman of the Advisory Committee on International Programs of the National Science Foundation, 1990-92. He has also served as consultant to the United Nations, Intergovernmental Oceanographic Commission of Unesco, Dept. of Fisheries of FAO, and the South Pacific Forum Fisheries Agency. In April 1993 he served as the UN-designated expert on GESAMP, the Joint Group of Experts on the Scientific Aspects of Marine Environmental Protection and in 1994 he was appointed Lead Author for Marine Policy in WG II-B (Oceans and Large Lakes) of the Intergovernmental Panel on Climate Change 1995, Re-assessment of the Global Climate Change Problem. Within the University of Washington, he has served as Director of the School of Marine Affairs (1982-1993), Chairman of the University Committee on Interdisciplinary Research and Graduate Education (1991-1992), and a member of the University's Steering Committee on Global Change (since 1992), and chairman of the President's Task Force on Environmental Education, 1995-1996. He was elected to membership in the NAS on April 29, 2003.

RAYMOND L. ORBACH is the Director of the Office of Science at the Department of Energy (DOE). As Director of the Office of Science (SC), Dr. Orbach manages an organization that is the third largest Federal sponsor of basic research in the United States which is viewed as one of the premier science organizations in the world. Prior to his appointment, Dr. Orbach served as Chancellor of the University of California at Riverside from April 1992 through March 2002; he now holds the title Chancellor Emeritus. Dr. Orbach began his academic career as a postdoctoral fellow at Oxford University in 1960 and became an assistant professor of applied physics at Harvard University in 1961. He joined the faculty of the University of California, Los Angeles (UCLA) two years later as an associate professor, and became a full professor in 1966. From 1982 to 1992, he served as the Provost of the College of Letters and Science at UCLA. Dr. Orbach has received numerous honors as a scholar including two Alfred P. Sloan Foundation Fellowships, a National Science Foundation Senior Postdoctoral Fellowship, a John Simon Guggenheim Memorial Foundation Fellowship, the Joliot Curie Professorship at the Ecole Superieure de Physique et Chimie Industrielle de la Ville de Paris, the Lorentz Professorship at the University of Leiden in the Netherlands, and the 1991-1992 Andrew Lawson Memorial Lecturer at UC Riverside. He is a fellow of the American Physical Society and the AAAS. Dr. Orbach received his B.S. degree in Physics from the California Institute of Technology in 1956. He received his Ph.D. degree in Physics from the University of California, Berkeley, in 1960 and was elected to Phi Beta Kappa.

JULIO DE PAULA is Professor of Chemistry and Director of the Marian E. Koshland Integrated Natural Sciences Center at Haverford. He is the recipient of the Henry Dreyfus Teacher-Scholar Award, a national honor bestowed on chemists who have excelled at both teaching and research. Funding for his research comes from the National Science Foundation. He has focused his years of research on the molecular interactions responsible for plant photosynthesis and on novel laser-based tumor treatments. He obtained his B.A. degree in Chemistry from Rutgers, The State University of New Jersey in 1982, and received a Ph.D. in Chemistry from Yale University in 1987. He was a recipient of an NIH Postdoctoral Fellowship in 1988 to conduct research at Michigan State University. He joined the Haverford faculty in 1989. Dr. Paula is the co-author of the Seventh Edition of "Physical Chemistry" with Peter Atkins, Oxford University.

MARIA PELLEGRINI joined the W. M. Keck Foundation as Program Director for Science, Engineering and Liberal Arts in February of 1998. She was Dean of Research in the College of Letters, Arts and Sciences at the University of Southern California from 1994 to 1998. Dr. Pellegrini was

Professor of Biological Sciences at USC from 1977 to 1998, serving as department chair from 1988 to 1993. She has taught a variety of courses in molecular biology and biochemistry at the undergraduate and graduate levels. Her research interests included studies of the structure-function relationships within ribosomes, the regulation of ribosomal gene expression, and, recently, work on genes that are important in human production. She has co-authored over 50 scientific journal articles and review chapters including an Institute for Scientific Information "citation classic." Dr. Pellegrini was the recipient of an Alfred P. Sloan Foundation Fellowship and a Dreyfus Foundation Teacher-Scholar Award. She has received numerous research and training grants from the National Institutes of Health. She has served on National Institutes of Health, California Breast Cancer Research Council and American Cancer Society grant review panels. She received her B.A. degree in chemistry from Connecticut College in 1969 and her Ph.D. in chemistry from Columbia University in 1974 followed by postdoctoral fellowships at Caltech and UC Irvine.

**FENIOSKY PEÑA-MORA** is currently an O'Neil Faculty Scholar and Associate Professor in the Department of Civil and Environmental Engineering at the University of Illinois at Urbana-Champaign. Peña-Mora was previously an Associate Professor of Information Technology and Project Management in the Civil and Environmental Engineering Department's Intelligent Engineering Systems Group at the Massachusetts Institute of Technology. His current research interests are in information technology support for collaboration, change management, conflict resolution, and process integration during design and development of large-scale civil engineering systems. He is the author of publications on computer-supported design, computer-supported engineering design and construction, project control and management of large-scale engineering systems. One of his publications received the 1995 award for best paper published in the ASCE Journal of Computing in Civil Engineering. Another of his publications is the textbook entitled "Introduction to Construction Dispute Resolution." He is also holder of a 1999 NSF CAREER Award and a 2000 White House PECASE (Presidential Early Career Award for Scientists and Engineers) Award. He is an Associate Editor for the ASCE Journal of Computing in Civil Engineering and the ASCE Journal of Construction Engineering and Management.

**DIANA RHOTEN** is a program office for the Social Science Research Council. She has a Ph.D. in Social Sciences, Policy, and Educational Practice and an M.A. in Organizational Sociology from Stanford University, as well as an M.Ed. in International Development Education from Harvard University. From 2001 to 2003, Dr. Rhoten served as an assistant professor at the Stanford University School of Education where she taught courses in international education development and interdisciplinary research methods. At this time, Dr. Rhoten was also the research director of the Hybrid Vigor Institute and the principal investigator of the Institute's NSF-funded study on interdisciplinary research networks and methods. In addition to analyzing interdisciplinary research organizations, Dr. Rhoten also studies cross-programmatic strategies in philanthropy.

**CATHERINE ROSS** is the Georgia Tech College of Architecture's first endowed faculty member—the Harry West Chair for Quality Growth and Regional Development. In this role, Dr. Ross directs a center that examines key issues of land use, community design, transportation and air quality throughout the Atlanta region and beyond. She grew up in Ohio, graduated from Kent State University, and received her Ph.D. in Urban and Regional Planning at Cornell University. She did post-doctorate work at the University of California at Berkeley. In addition, Ross founded a consulting company that has conducted research for numerous government transportation agencies, and has been published extensively in the fields of urban planning, transportation planning and public participation. Dr. Ross has served as senior policy advisor at the National Academy of Sciences Transportation Research Board and vice provost for academic affairs at Georgia Tech. She is past president of the National Association of Collegiate Schools of Planning and was recently appointed to the national advisory board of the Women's Transportation Seminar. She also serves as vice chair of the Atlanta Development Authority.

**F. SHERWOOD ROWLAND** is a specialist in atmospheric chemistry and radiochemistry, and was, with colleague Mario Molina, the first scientist to warn that chlorofluorocarbons (CFCs) released into the atmosphere were depleting the earth's critical ozone layer. Dr. Rowland arrived at the University of California, Irvine, in 1964 as the first chair of the Department of Chemistry. He previously held faculty positions at Princeton University and the University of Kansas. He holds a bachelor's degree from Ohio Wesleyan University, a master's and a doctorate from the University of Chicago, and a number of honorary degrees from universities in the United States and the United Kingdom. Rowland is a member of the National Academy of Sciences and the American Academy of Arts and Sciences. During 1991–1993, he served successive one-year terms as President-Elect, President, and Chairman of the Board of the American Association for the Advancement of Science. Dr. Rowland was awarded the American Chemical Society 1993 Peter Debye Medal in Physical Chemistry, and the 1994 Roger Revelle Medal from the American Geophysical Union. In 1995, he shared the Nobel Prize in Chemistry with Mario Molina and Paul Crutzen.

**LAWRENCE A. TABAK** is the director of the National Institute of Dental and Craniofacial Research (NIDCR). The former director of the Center for Oral Biology, Aab Institute, at the University of Rochester in New York, Dr. Tabak also served as senior associate dean for research at the School of Medicine and Dentistry. While at Rochester, he oversaw a number of interdisciplinary research groups studying the molecular and genetic aspects of craniofacial-oral-dental conditions. He also directed graduate research training programs at the university and held professorships in dentistry and biochemistry and biophysics. Dr. Tabak has also served in various official capacities in a number of professional organizations, including the International/American Association for Dental Research, the American Association for the Advancement of Science, and the Society for Glycobiology. He has received numerous honors and awards for his work, including being named a fellow of the AAAS and most recently, his election to the Institute of Medicine of the National Academies. A native of Brooklyn, New York, Dr. Tabak received his undergraduate degree from City College of the City University of New York, his D.D.S. from Columbia University, and both a Ph.D. and certificate of proficiency in endodontics from the State University of New York at Buffalo.

**JEFFREY WADSWORTH** is the director of Oak Ridge National Laboratory, the largest multipurpose laboratory of the U.S. Department of Energy (DOE), with 3,800 staff members and an annual budget of $1 billion. He is also a corporate officer of Battelle Memorial Institute, in Columbus, Ohio, where he is senior vice president for DOE Science Programs. He joined Battelle in August 2002 and was a member of the White House Transition Planning Office for the U.S. Department of Homeland Security. He previously served as Deputy Director for Science and Technology at DOE's Lawrence Livermore National Laboratory, as well as Associate Director for Chemistry and Materials Science at that laboratory. Dr. Wadsworth holds B.S., Ph.D., and D. Met. degrees in metallurgy from the University of Sheffield in England. He is a Fellow of the American Society for Metals and the Minerals, Metals, and Materials Society. In 2003, he was elected a Fellow of the American Association for the Advancement of Science in recognition of "distinguished contributions in developing advanced materials and superplasticity, and in determining the history and origins of Damascus and other steels, and for broad scientific leadership supporting national security."

**PIERRE WILTZIUS** is director of the Beckman Institute for Advanced Science and Technology; a professor in both the Department of Materials Science and Engineering and the Department of Physics; and a full-time Beckman Institute faculty member in the Nanoelectronics and Biophotonics

Group. His fields of professional interest are soft-condensed matter, colloidal self-assembly, photonic crystals and microphotonics. Pierre Wiltzius received his Ph.D. in physics from the Swiss Federal Institute of Technology (ETHZ), Zurich, Switzerland in 1981. He was at Bell Laboratories (Lucent Technologies, formerly AT&T) between 1984 and 2001, where he was most recently the Director of Semiconductor Physics Research. He is a Fellow of the American Physical Society; a Fellow of the American Association for the Advancement of Science; a Senior Member of the IEEE; and a recipient of a NATO Fellowship. Interdisciplinary research has been central to his professional career. His Ph.D. thesis was on aspects of blood coagulation and was the result of a collaboration between physicists and clinical physicians.

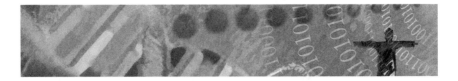

## REPRESENTED ORGANIZATIONS

The following organizations are represented at the Convocation on Facilitating Interdisciplinary Research.

Aerospace Corporation
Abt Associates, Inc.
Alliance for Academic Internal Medicine
American Association of Colleges of Pharmacy
American Chemical Society
American College of Radiology
American Health Information Management Association
American Institute of Biological Sciences
American Institute of Physics
American Mathematical Society
American Museum of Natural History
American Psychological Association
American Psychological Society
American Society of Cell Biology
American Society of Plant Biologists
American Sociological Association
Arizona State University
Arnold & Porter
ASHP Research & Education Foundation
Association of American Geographers
Atlantic Philanthropies (USA)
Baltimore City Public Schools
Bar-Ilan University
BART IGERT: Biosphere-Atmosphere Research and Training Program
Biophysical Society
Brookhaven National Laboratory
Buffalo State College
Burroughs Wellcome Fund
California State University Program for Education and Research in
    Biotechnology
Canadian Institute for Advanced Research
Consortium for Oceanographic Research and Education

Contemporary Communications, Inc.
Cornell University
Council on Undergraduate Research
Des Moines University
Duke Center for Environmental Solutions
Duke University
Ecological Society of America
Embassy of France
Embassy of Switzerland
EnTech Strategies, LLC
Environment Canada
Experimental Program to Stimulate Experimental Research Foundation
Faculty Career and Diversity Consultant
Federation of American Societies for Experimental Biology
Flattau Associates, LLC
Florida A & M University
Food and Drug Administration
Fred Hutchinson Cancer Research Center
George Washington University
Georgia Institute of Technology
Graduate Partnerships Program
Graduate School of Public Health, University of Pittsburgh
Gulf Coast Consortia
Harvard University
Health Resources and Services Administration
House Resources Committee
Howard Hughes Medical Institute
Howard University
Independent Consultant
Industry-University Cooperative Research Program
Institute for Prevention Research
James Madison University
JMW Associates
John Templeton Foundation
Johns Hopkins University, Bloomberg School of Public Health
Land Information and Computer Graphics Facility
Lewis-Burke Associates
Lincoln University of Pennsylvania
Louisiana Tech University
Mathematical Association of America
McGeary and Smith
Medical College of Georgia
Michigan State University

Montclair State University
Mouvement Burkinabè d'Ecologie
National Aeronautics Space Administration Marshall Space Flight Center
National Cancer Institute
National Education Knowledge Industry Association
National Institute of Biomedical Imaging and Bioengineering
National Institute of Mental Health
National Institute of Standards and Technology
National Institute on Drug Abuse
National Institutes of Health
National Institute on Disability and Rehabilitation Research
Natural Sciences and Engineering Research Council of Canada
National Institute for Biomedical Imaging and Bioengineering/National
    Institutes of Health
National Institute of General Medical Sciences
National Institutes of Health/National Heart, Lung, & Blood Institute
National Institute of Mental Health
National Oceanic and Atmospheric Administration/OAR
Northern Arizona University
Northwestern University
National Science Foundation
Ohio Agricultural Research and Development Center/Ohio State
    University
Ohio State University
Office of Management and Budget
Office of Naval Research Global
Office of Science & Technology Policy
Office of the Director, NIH
Office of Translational Research & Scientific Technology
Oklahoma State University
Orthotic and Prosthetic Assistance Fund, Inc.
Paralyzed Veterans of America
Pennsylvania State University
Potomac College
Purdue University
Research for Better Schools
Rutgers University
Sandia National Laboratories
SETI Institute
Social Science Research Council
Society for Women's Health Research
Southern Illinois University Carbondale
Stony Brook University

Syracuse University
Technology Policy and Assessment Center
TechVision21
Texas Tech University
Thomas Jefferson University
University of Maryland—Baltimore Campus
UMDNJ-Robert Wood Johnson Medical School
Uniformed Services University
University of Nebraska Medical Center
University at Buffalo
University of California
University of California, Los Angeles
University of California, Davis
University of California, Irvine
University of Cincinnati
University of Colorado
University of Florida
University of Georgia
University of Kansas
University of Massachusetts, Lowell
University of Michigan
University of Nebraska, Lincoln
University of New Mexico
University of North Carolina at Chapel Hill
University of North Carolina at Greensboro
University of North Dakota
University of Oklahoma, Tulsa Graduate College
University of Oregon
University of Pittsburgh
University of the Philippines Baguio
University of Wisconsin, Milwaukee
U.S. Department of Agriculture—Cooperative State Research, Education, and Extension Service
U.S. Department of Commerce
U.S. Department of Energy
U.S. Environmental Protection Agency
U.S. State Department
Utah Addiction Center
University of Texas at Dallas
University of Texas Medical Branch
Vanderbilt University
Vanderbilt University Medical Center
Virginia Polytechnic Institute and State University

W. M. Keck Foundation
Washington State University
Washington University School of Medicine
Water Environment Research Foundation
Wind Hollow Foundation
Women in International Security
Yale University School of Medicine

# D

# From Interdiscipline to Discipline

The relationship between interdisciplinary and disciplinary research is dynamic. Researchers in one discipline may follow a question to the interface of another discipline and return "home" with new knowledge. If the journey is especially productive, it may cross one or more intellectual frontiers to produce a new discipline.

As discussed in Chapter 2, this process of interdisciplinarity has been propelled by a number of "drivers." For example, the driver of generative technologies may be said to have given rise to partnerships between biology and chemistry more than two centuries ago after Lavoisier's studies of combustion and Priestley's discovery of the presence of oxygen in the air. And the partnerships coalesced over the years in the new "interdiscipline" of biochemistry, which emerged with its own distinctive character and is now generally considered a discipline.

In most cases, emerging disciplines become mature when they attract a critical mass of participants whose increasing numbers and productivity warrant a new set of societies, journals, and academic departments. The founders of the distinct discipline, who were usually trained in one of its "parent" disciplines, may then take the logical, although often discomfiting step, of moving into a new professional identity and culture.

The purpose of this appendix is to illustrate, by example, how interdisciplinary partnerships have evolved into new disciplines and how these new disciplines have led to the creation of a new breed of interdisciplinary professional society since World War II. This issue is discussed further in Chapter 7 on the role of professional societies in interdisciplinary research.

## GEOBIOLOGY

The recent emergence of geobiology into a mature field was preceded by a long gestation period, beginning with the pioneering studies of the earth's surface by James Hutton more than two centuries ago. By the beginning of the 20th century, the great Russian polymath Vladimir Vernadsky focused more explicitly on the influence of the biosphere (including human activities) on geological processes, and the term *geobiology* was first used soon afterward by the Dutch biologist Lourens Bass Becking in 1934. Most recently, the extensive writings of the independent scientist James Lovelock served to highlight the role of life in influencing the surface environment of the earth.[1]

Awareness of the importance of geobiology was widened by technologies that revealed new kinds of organisms that flourish in remote and extreme environments. Discoveries of how microbes contribute to geochemical reactions or react with the geosphere in novel ways have stirred the excitement of many who seek solutions to a wide array of environmental and resource challenges. Among the existing disciplines that have fed the growth of geobiology are geochemistry, geohydrology, oceanography, microbiology, environmental studies, biogeochemistry, ecology, molecular biology, genomics, paleobiology, and mineralogy.

The interaction of biological and geological thinking developed over many decades, but the formal birth of the new field happened quickly. It was stimulated in part by the report of a colloquium held in December 2000 by the American Academy of Microbiology, which formally described geobiology as "research that attempts to understand the interface between the biosphere and the geosphere." The report was followed by the decision of the Geological Society of America to create the new Geobiology and Geomicrobiology Division in May 2001 and then by the decisions of Elsevier Science to publish *Virtual Journal of Geobiology* in 2002 and of Blackwell Publishing to launch the new journal *Geobiology* in 2003. The University of Southern California Wrigley Institute for Environmental Studies held an "International Training Course in a Rapidly Evolving Field: Geobiology" in June 2004.[2]

---

[1]Lovelock's assertion that the "planet Gaia" is a "self-regulating" system has stirred controversy, but his elucidation of biosphere-geosphere interactions is nonetheless extensive.

[2]See the colloquium report "Geobiology: Exploring the Interface Between the Biosphere and the Geosphere, 2001, at *http://www.asm.org/Academy/index.asp?bid=2132.*

## NEUROSCIENCE

Neuroscience has been defined as the interdisciplinary investigation of the nervous system and behavior.[3] Thomas Willis, an English anatomist, provided the first detailed description of brain structure in the middle 1600s, and 200 years later scientists began to correlate structures with functions. By the end of the 19th century, brain research institutes began to formalize research activity in the structure of universities.

Until a few decades ago, most scientists engaged in brain research identified themselves with anatomy, physiology, psychology, biochemistry, and other disciplines. Then, in the 1960s, a "critical mass" of brain researchers around the world felt the need to focus their activities on a single framework and to formalize neuroscience as a discipline. In response, the International Brain Research Organization was founded in 1960 to promote cooperation among the world's scientific resources for research on the brain. The British Brain Research Association was founded in 1968; it is now the British Neuroscience Association. In the United States, the Society for Neuroscience was founded in 1969, with its official organ, the *Journal of Neuroscience*. Membership in the US society grew from 1,000 in 1970 to about 34,000 in 2000.

In this new discipline, neuroscientists are integrating a variety of perspectives to gain insights into fundamental questions about the nervous system in health and disease. According to a recent study, "Neuroscience is a clear example of a discipline of today arising from interdisciplinary approaches of the past."[4] Like other emerging fields, it interacts with other disciplines and techniques as needed, including informatics and molecular biology. It has been invigorated by new technologies, such as the use of positron emission tomography to image blood flow and magnetic resonance imaging to look at neural structures. Its growth has been so rapid that some of its own subdisciplines, such as cognitive neuroscience, are now acquiring disciplinary status.

### SUSTAINABILITY SCIENCE AND ENGINEERING

In contrast with the previous two examples, the concept behind sustainability science is relatively young, having evolved largely out of the environ-

---

[3]Frank, R. J., Marshall, L. H., and Magoun, H. W. "The Neurosciences," In Bowers, J. Z. and Purcell, E. F., *Advances in American Medicine: Essays at the Bicentennial, Vol. 2,* Josiah Macy Jr. Foundation, 1976.

[4]Institute of Medicine, *Bridging Disciplines in the Brain, Behavioral, and Clinical Sciences,* Washington, D.C.: National Academy Press, 2000.

mental movement of the 1960s and 1970s. That decade saw growth in the awareness of a linked series of environmental problems, including resource depletion, population growth, and pollution of air, water, and soil.

Initially, environmental studies focused on issues of waste management, especially on air, water, and soil pollution. The strategy for treating pollutants focused on "end-of-pipe" techniques and other local measures. As it became clear that end-of-pipe measures were merely palliative, they evolved toward the broader activities of pollution prevention, conservation, and social policies.

By 1987, a report from the UN-mandated Brundtland Commission could describe "sustainable development" as development "which meets the needs of the present without compromising the ability of the future to meet its needs."[5] That report served as a catalyst for the 1992 UN Conference on Environment and Development (the "Earth Summit") in Rio de Janeiro. The evidence delivered at the conference made it clear that it was necessary "to integrate the physical and social science disciplines with engineering to address the ecological, economic, social, and political processes that determine the sustainability of natural and human life cycles and activities."[6] Thus arose the need to develop an interdisciplinary infrastructure, termed *sustainability science and engineering*. The broad goals of this field are to define major threats to sustainability, find accurate indicators of change (from children's birth weights to atmospheric chemistry), and explore promising opportunities for circumventing or mitigating environmental threats.

Although it may be premature to define this field as a stand-alone discipline,[7] some researchers have articulated a vision of a "metadiscipline." For example, one paper defines sustainability as "the design of human and industrial systems to ensure that humankind's use of natural resources and cycles do not lead to diminished quality of life due either to losses in future economic opportunities or to adverse impacts on social conditions, human health, and the environment."[8] It remains to be seen whether an enterprise of such breadth is a discipline in the traditional sense or whether researchers are leading us toward a new concept of the discipline.

---

[5]World Commission on Environment and Development, *Our Common Future*, New York: Oxford University Press, 1987.

[6]National Research Council, *Our Common Journey: A Transition Toward Sustainability*, 1999.

[7]Clark, W. C. and Dickson, N. M. "Sustainability science: The emerging research program," *Proceedings of the National Academy of Sciences*, 100(14):806, 2003.

[8]Mihelcic, J. R. et al., "Sustainability Science and Engineering: The Emergence of a New Metadiscipline," *Environmental Science and Technology* 37(23):5315, 2003.

## CONCLUSION

Perhaps the most common driver of interdisciplinarity toward the emergence of new disciplines is the sheer complexity of nature, which draws researchers toward the next important question, moving toward interfaces with other disciplines and partnerships with colleagues in them. In the three examples above, the intellectual journey seems to be natural and even inevitable for those seeking answers to the questions of science and engineering. The more institutions and funding organizations can help these pioneer investigators along their way, the greater the intellectual and practical rewards of research are likely to be.

# E

# Survey of Institutions and Individuals Conducting Interdisciplinary Research

To enhance scholarship and collect quantitative data on the impediments, programs, and evaluation criteria related to interdisciplinary research (IDR), the committee developed survey instruments and disseminated them to provosts and others.[1] In this appendix, we analyze the results of the committee's surveys of those interested in IDR, including students, postdoctoral scholars, faculty, funders, policy makers, and disciplinary societies.

The first survey, referred to in the report as the "convocation survey," was given to the 150 persons who attended the Convocation on Facilitating Interdisciplinary Research, on January 29-30, 2004 (see Appendix C); 91 convocation participants responded to the survey—about a 75 percent return rate. The "individual survey," a slightly modified version of the convocation survey, was posted on the committee's Web site. An invitation to participate in the survey was sent to universities, professional societies, nongovernment organizations, and participants in federal and private interdisciplinary programs; 423 people responded to the solicitation. An invitation to participate in a third survey, the "provost survey," was distributed on line to provosts or vice-chancellors of institutions that conduct IDR; 57 institutions responded.

---

[1] *http://www7.nationalacademies.org/interdisciplinary/SurveyHome.html.* The survey instrument for individuals is appended. It differs from the provost survey in question #1.

It must be noted that the survey population does not represent a random sample. There was undoubtedly selection bias in those who attended the Convocation on Facilitating Interdisciplinary Research and in those who responded to the Web-based survey. The results are representative of a wide population of researchers, but cannot be extrapolated to the entire population of researchers involved in interdisciplinary projects and programs. That said, the findings corroborate and extend previous studies of IDR, and offer unique insights on joint appointments and differences between researchers and administrators, and provide suggestions for how to prioritize change efforts.

## DISSEMINATION

The convocation survey was distributed at the convocation in Washington, D.C. and the individual survey was distributed by the following organizations. We made every attempt to distribute the survey as widely as possible. Our strategy was to request larger organizations and umbrella societies in a variety of fields to distribute the survey

- American Chemical Society (ACS)
- American Institute of Biological Sciences (AIBS)
- Association for Integrative Studies
- Association of American Medical Colleges (AAMC)
- Association of American Universities (AAU)
- Association of Independent Research Institutes
- Biophysical Society
- Council of Graduate Students (CGS)
- Federation of American Societies for Experimental Biology (FASEB)
- National Association of State Universities and Land-Grant Colleges (NASULGC)
- National Academy of Public Administration
- National Institutes of Health Bioengineering Consortium (NIH BECON)
- DOE National Laboratories
- National Science Foundation (NSF) Engineering Research Centers
- NSF Frontiers in Integrative Biological Research (FIBR) awardees
- NSF Integrative Graduate Education and Research Traineeships (IGERT) awardees
- NSF Science and Technology Centers
- Washington Science Policy Alliance

The following institutions participated in the provost survey. We received the assistance of NASULGC and AAU in distributing the survey to their member universities.

- Barnard College
- Boston University
- Carnegie Mellon University
- Cedars-Sinai Medical Center
- Clarkson University
- Columbia University
- Department of Energy Idaho Operations Office
- Florida State University
- Georgia State University
- Instituto Mexicano del Seguro Social
- Iowa State University
- Jackson Laboratory
- Johns Hopkins University
- Lewis & Clark College
- Massachusetts Institute of Technology
- Medical College of Georgia
- Miami University
- National Cancer Institute
- National Dairy Council
- New York University
- North Dakota State University
- Northwestern University
- Pennsylvania State University
- Purdue University
- Simon Fraser University
- Stanford University
- Syracuse University
- Texas A&M University
- Tulane University
- University at Buffalo
- University of Arizona
- University of California, Irvine
- University of California, Los Angeles
- University of California, Santa Barbara
- University of Chicago
- University of Cincinnati College of Medicine
- University of Houston
- University of Idaho
- University of Illinois, Chicago

- University of Maryland, Baltimore County
- University of Michigan
- University of Minnesota
- University of Missouri, Columbia
- University of North Carolina, Chapel Hill
- University of Tennessee
- University of Texas, Austin
- University of Utah
- University of Washington
- Vanderbilt University
- Virginia Polytechnic Institute and State University
- Wayne State University
- Wright State University

## SURVEY DEMOGRAPHICS

The committee collected information on respondent position and rank, involvement in IDR, age, and institution type, size, and budget.

### Position and Rank

Respondents were predominantly faculty researchers, administrators, or both.

| Position | Convocation n | % | Individual n | % | Provost n | % |
|---|---|---|---|---|---|---|
| Student | 2 | 2.2 | 26 | 6.2 | 0 | 0 |
| Postdoctoral scholar | 0 | 0.0 | 18 | 4.3 | 0 | 0 |
| Researcher/faculty | 29 | 31.9 | 325 | 76.8 | 3 | 5.3 |
| Administrator | 26 | 28.6 | 5 | 1.2 | 12 | 21.1 |
| Researcher/admin. | 17 | 18.7 | 47 | 11.1 | 40 | 70.2 |
| Funder | 16 | 17.6 | 0 | 0 | 0 | 0 |
| Other/not answered | 1 | 1.1 | 2 | 0.5 | 2 | 3.5 |
| Total | 91 | 100.1 | 423 | 100 | 57 | 100.1 |

Respondents to the convocation and provost surveys predominantly held senior positions. The individual survey showed a wider array of ranks, but people holding senior-level positions outnumbered middle-level and junior positions by 2 to 1.

| Rank | Convocation | | Individual | | Provost | |
|---|---|---|---|---|---|---|
| | n | % | n | % | n | % |
| Senior | 64 | 70.3 | 194 | 45.9 | 52 | 91.2 |
| Middle-level | 17 | 18.7 | 113 | 26.7 | 1 | 1.8 |
| Junior | 6 | 6.6 | 105 | 24.8 | 2 | 3.5 |
| Not answered | 4 | 4.4 | 11 | 2.6 | 2 | 3.5 |
| Total | 91 | 100.0 | 423 | 100.0 | 57 | 100.0 |

## Age Distribution

Overall, age distribution was fairly normal, with a mean of about 50 years.

| Age | Convocation | | Individual | | Provost | | Total | |
|---|---|---|---|---|---|---|---|---|
| | n | % | n | % | n | % | n | % |
| 20-29 | 3 | 3.3 | 31 | 7.3 | 0 | | 34 | 6.0 |
| 30-39 | 11 | 12.1 | 103 | 24.3 | 1 | 1.8 | 115 | 20.1 |
| 40-49 | 27 | 29.7 | 122 | 28.8 | 7 | 12.3 | 156 | 27.3 |
| 50-59 | 35 | 38.5 | 95 | 22.5 | 30 | 52.6 | 160 | 28.0 |
| 60-69 | 11 | 12.1 | 48 | 11.3 | 12 | 21.1 | 71 | 12.4 |
| >70 | 3 | 3.3 | 6 | 1.4 | 0 | | 9 | 1.6 |
| Not answered | 1 | 1.1 | 18 | 4.3 | 7 | 12.3 | 26 | 4.6 |
| Total | 91 | 100.1 | 423 | 99.9 | 57 | 100.1 | 571 | 100.0 |

## Type of Institution

The majority of respondents were working at public academic institutions. About half as many worked at private academic institutions. (See Figure E-1.) Industry researchers, funders, and disciplinary-society representatives were targeted for participation only at the convocation and are not represented in the individual or provost survey populations.

| Type of Institution | Convocation | | Individual | | Provost | |
|---|---|---|---|---|---|---|
| | n | % | n | % | n | % |
| Public academic | 42 | 46.2 | 264 | 62.4 | 33 | 57.9 |
| Private academic | 15 | 16.5 | 122 | 28.8 | 17 | 29.8 |
| Industrial R&D org. | 2 | 2.2 | 3 | 0.7 | 0 | |
| Government R&D org. | 3 | 3.3 | 17 | 4.0 | 3 | 5.3 |
| Indep. research inst. | 3 | 3.3 | 9 | 2.1 | 1 | 1.8 |
| Public funding inst. | 9 | 9.9 | 8 | 1.9 | 0 | |
| Private funding inst. | 8 | 8.8 | 0 | | 0 | |
| Professional society | 6 | 6.6 | 0 | | 0 | |
| Other/not answered | 8 | 8.8 | 0 | | 3 | 5.3 |
| Total Surveys (Total[a]) | 91(96) | 105.6 | 423 | 101.8 | 57 | 100.1 |

[a]Some respondents gave multiple answers to this question. Percent is calculated using the total number of surveys returned, and may add up to more than 100%.

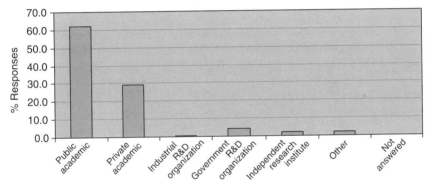

Type of Institution

**FIGURE E-1**   Type of institutions responding.

## Size, Budget, and Number of Researchers

Survey respondents were asked to indicate the annual budget of their institutions and the numbers of faculty, undergraduates, graduate students, and postdoctoral fellows (see Figure E-2). It appears that most respondents were working at large research institutions. Annual budgets showed a bimodal distribution, with peaks at $10 million–100 million and over $1 billion. At the same time, almost half the respondents indicated that they

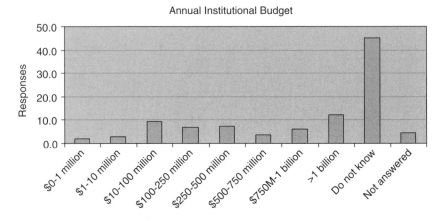

**FIGURE E-2**   Annual institutional budgets.

were not aware of their institutions' annual budget. Responses indicated that institutions tended to have over 500 faculty, 10,000 undergraduates, and over 2,500 graduate students (Figures E-3, E-4, and E-5). Most respondents did not know how many postdoctoral fellows were at their institutions (Figure E-6).

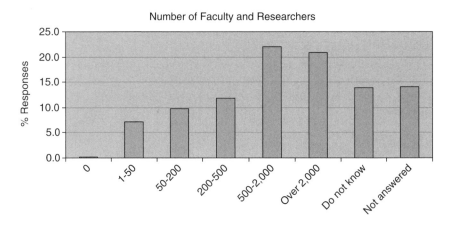

**FIGURE E-3**   Number of faculty and researchers at the respondents' institutions.

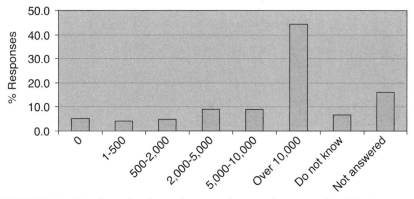

**FIGURE E-4**   Number of undergraduate students at the respondents' institutions.

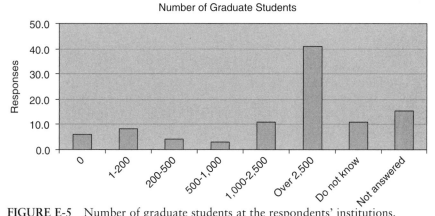

**FIGURE E-5**   Number of graduate students at the respondents' institutions.

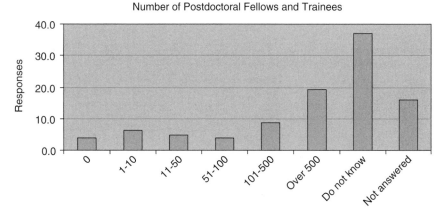

**FIGURE E-6**   Number of postdoctoral fellows and trainees at the respondents' institutions.

## RELATIONSHIP TO INTERDISCIPLINARY RESEARCH

### Participation in Interdisciplinary Research

In the combined surveys, 94 percent of respondents were at least partially involved in IDR.

| Participation | Convocation | | Individual | | Provost | |
|---|---|---|---|---|---|---|
| | n | % | n | % | n | % |
| Primarily interdisciplinary | 53 | 58.2 | 263 | 62.2 | 24 | 42.1 |
| Partially interdisciplinary | 28 | 30.8 | 147 | 34.8 | 22 | 38.6 |
| Not interdisciplinary | 0 | | 12 | 2.8 | 4 | 7.0 |
| Not answered | 10 | 11.0 | 1 | 0.2 | 7 | 12.3 |
| Total | | 100.0 | 433 | 102.4 | 57 | 100.0 |

### Specific Roles

Respondents were asked to indicate how they were involved in IDR. This was a free-answer section; responses were analyzed and categorized by staff. Because more than one answer could have been provided, results may add up to more than 100 percent.

| Involvement in IDR | Convocation | | Individual | | Provost | |
|---|---|---|---|---|---|---|
| | n | % | n | % | n | % |
| Oversee or support IDR programs | 19 | 23.5 | 0 | 0 | 45 | 97.8 |
| Fund IDR programs or grants | 14 | 17.3 | 0 | 0 | | |
| Research is interdisciplinary | 41 | 50.6 | 366 | 89.3 | 23 | 50.0 |
| Collaborate with others in different disciplines | 3 | 3.7 | 97 | 23.7 | 2 | 4.3 |
| Head/director of IDR program | 7 | 8.6 | 28 | 6.8 | 1 | 2.2 |
| Involved with IDR training program or teach IDR classes | 2 | 2.5 | 18 | 4.4 | 1 | 2.2 |
| Editor of IDR journal | 0 | 0.0 | 2 | 0.5 | 0 | 0 |
| Other | 8 | 9.9 | 8 | 2.0 | 0 | 0 |
| Total involved in IDR | 81 | | 410 | | 46 | |
| Not interdisciplinary or not answered | 10 | | 13 | | 11 | |

## Ranking of Institutional Environment for IDR

Respondents were asked to rank the general supportiveness for IDR at their current institution and up to two previous institutions on a scale of 0 (IDR-hostile) to 10 (IDR-supportive). There appears to be a trend toward more supportive environments for IDR. It is possible that respondents moved to institutions that were more supportive during the course of their careers. Rankings are reported as mean +/– standard deviation. Not all respondents provided an answer. The total number of responses to this question was n = 480.

| Environment for IDR | Convocation | Individual | Provost |
|---|---|---|---|
| Current institution | 7.74 +/– 2.07 | 7.25 +/– 2.31 | 7.24 +/– 1.70 |
| Previous institutions | 5.95 +/– 2.17 | 6.35 +/– 2.57 | 5.67 +/– 2.04 |

To determine whether rank was associated with institution size or budget, we sorted the rankings by annual budget, number of faculty, and number of undergraduates (see Figures E-7 and E-8). There was no relationship between number of undergraduates and ranking, but there are some interesting trends for budget and number of faculty. It appears that smaller or larger institutions have a better environment for IDR than those with intermediate budget and faculty numbers.

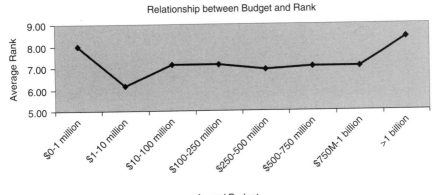

FIGURE E-7   Relationship between institutional budget and rank.

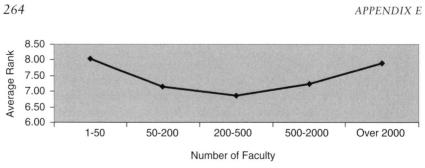

**FIGURE E-8** Relationship between number of faculty and rank.

## INTERDISCIPLINARY RESEARCH AT INSTITUTIONS

When asked whether there were impediments to IDR at their current institutions, 70.7 percent of the respondents answered yes, 23.2 percent answered no, and 6.2 percent did not know or did not answer (see Figure E-9).

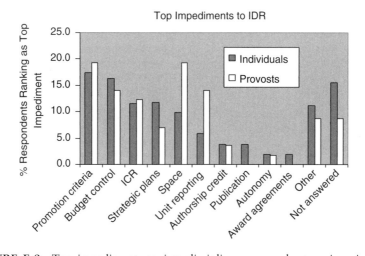

**FIGURE E-9** Top impediments to interdisciplinary research at various institutions.

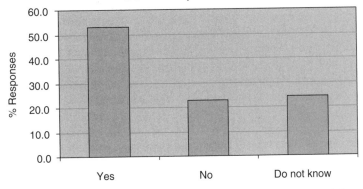

FIGURE E-10   Survey results as to whether seed money was provided for IDR.

Respondents were provided a list and asked to rank the top five impediments to IDR at their institutions (see Figure 4-5). The list[2] included budget control, indirect-cost recovery (ICR), publication in disciplinary and interdisciplinary journals, compatibility with college or department strategic plans, promotion and tenure criteria, credit for joint authorship, unit reporting relationships, space allocation, honoring award agreements, restrictions on faculty autonomy, and other. The chart indicates the percentage of respondents who gave an impediment top ranking. It is interesting to note that "individuals" and provosts ranked impediments differently. Furthermore, impediments often mentioned in research literature–authorship credit and publication–were among the lowest ranked by both respondent groups. The impediments that were most often ranked first by "individuals" were promotion criteria, budget control, ICR, and compatibility with strategic plans. For provosts, the top impediments were promotion criteria, space allocation, budget, and ICR.

## Seed Money

Respondents were asked whether their institution provided seed money to help start up interdisciplinary programs and were asked to briefly describe the amounts available and the major criteria used in making awards. Over half the institutions provided such "venture capital" for interdisciplinary work. Amounts provided ranged from $1,000 to $1 million. Duration of awards also varied but tended to be short: 1- to 2-year grants (see Figures E-10, E-11, and E-12).

---

[2]Feller, I. "New Organizations, old culture. Strategy and Implementation of Interdisciplinary Programs." AAAS Annual Meeting Presentation. February 16, 2002.

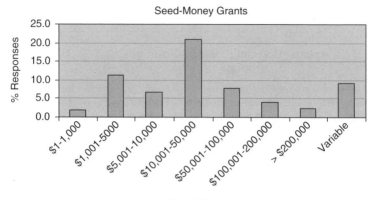

Size of Award

**FIGURE E-11** Seed money grants and the size of the award.

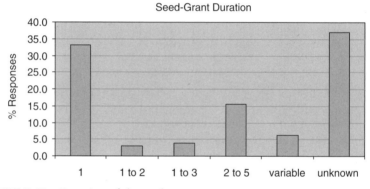

**FIGURE E-12** Duration of the seed grant.

Three main criteria were cited by survey respondents for evaluating proposals for seed money:

1.  What is the likelihood that this project or program, once developed, would generate outside funding? (21 percent)
2.  What is the scientific merit of the work? (20 percent)
3.  Is the work truly interdisciplinary? (20 percent)

"Other" responses (19.8 percent) ranged from selection-committee biases to university or department long-term strategic goals. Respondents often cited more than one criterion for determining seed-money allocation; therefore, the percentage of responses (based on the number of respondents) exceeds 100 percent (see Figure E-13).

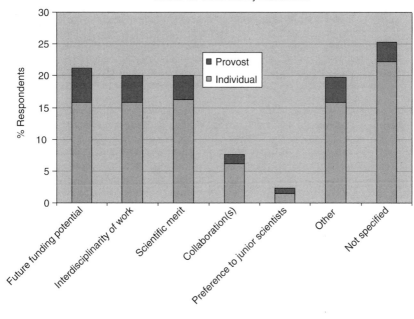

FIGURE E-13   Criteria for seed-money distribution.

## Joint Appointments

When asked whether their institutions made joint appointments for interdisciplinary faculty or staff in which salary is shared, most respondents answered yes. However, in most cases, fewer than 10 percent of the faculty at the respondents' institutions held such joint positions.

| Shared Salary for Joint Appointments? | Individual | | Provost | |
|---|---|---|---|---|
| | n | % | n | % |
| Yes | 249 | 58.9 | 42 | 73.7 |
| No | 85 | 20.1 | 12 | 21.0 |
| Do not know | 88 | 20.8 | 2 | 3.5 |
| Not answered | 1 | 0.2 | 1 | 1.8 |
| Total | 423 | 100.0 | 57 | 100.0 |

| Proportion of Such Joint Appointments | Individual | | Provost | |
|---|---|---|---|---|
| | n | % | n | % |
| 0-10% | 190 | 76.3 | 33 | 78.6 |
| 10-25% | 24 | 9.6 | 7 | 16.7 |
| Over 25% | 6 | 2.4 | 1 | 2.4 |
| Did not answer | 29 | 11.6 | 1 | 2.4 |
| Total (based on those who answered yes above) | 249 | 100 | 42 | 100.1 |

## Interdisciplinary Programs and Characteristics

Respondents were asked to list and describe up to three interdisciplinary programs at their institutions with which they were currently involved, including centers and teaching programs. They were asked to indicate the number and name of each involved department, whether extra-institutional groups were involved, the number of researchers, whether there were associated faculty lines or training slots, the sources of funding, whether there was a central facility for the program, and how space was allocated. Over 800 programs were described, and this yielded rich data for anyone interested in examining the current organizational structure of IDR programs and centers. Among the findings, respondents indicated that 29.5 percent of the centers and programs did have faculty lines, whereas 33.3 percent did not; 12.3 percent stated that faculty lines did not apply to the program listed, and 24.7 did not know or did not provide an answer. The percentage of associated training slots was higher: 40.9 percent of programs listed had such slots, 23.1 percent did not.

## EVALUATION OF INTERDISCIPLINARY RESEARCH PROGRAMS

Respondents were asked to describe dominant forms of evaluation used by their institutions to evaluate interdisciplinary programs. The predominant methods of evaluation were internal and external visiting committees and informal feedback. Percentages add up to more than 100 because respondents could choose more than one answer.

| Institutional Evaluation Methods | Individual | | Provost | |
|---|---|---|---|---|
| | n | % | n | % |
| Internal committee | 148 | 35.0 | 38 | 66.7 |
| Visiting committee | 130 | 30.7 | 46 | 80.7 |
| Informal feedback | 122 | 28.8 | 30 | 52.6 |
| Principal-investigator assessment | 113 | 26.7 | 24 | 42.1 |
| Interviews | 25 | 5.9 | 7 | 12.3 |
| Benchmarking surveys | 20 | 4.7 | 10 | 17.5 |
| Do not know | 155 | 36.6 | 1 | 1.8 |
| Other | 24 | 5.7 | 6 | 10.5 |
| Not answered | 35 | 8.3 | 3 | 5.3 |
| Total answers | 423 | | 57 | |

Respondents were also asked to report the top three methods that they used to evaluate the success of interdisciplinary programs. Respondents were provided a list and the opportunity to enter other options. The predominant IDR evaluation methods varied between individual researchers and provosts. For both groups, the top two choices were potential for innovation and increasing institutional funding. Provosts ranked enhancing the reputation of their institutions third, and individual researchers ranked enhancing student experiences third.

| Personal Evaluation Methods | Individual | | Provost | |
|---|---|---|---|---|
| | n | % | n | % |
| Level of (or potential for) scientific discovery or innovation | 239 | 56.5 | 46 | 80.7 |
| Increasing institution's research funding | 156 | 36.9 | 33 | 57.9 |
| Enhancing richness of undergraduate or graduate experience | 150 | 35.5 | 22 | 38.6 |
| Enhancing institution's reputation | 132 | 31.2 | 25 | 43.9 |
| Increasing ability to attract outstanding faculty or postdoctoral scholars | 123 | 29.1 | 28 | 49.1 |
| Societal relevance of problem being addressed | 97 | 22.9 | 15 | 26.3 |
| Quality of leadership | 95 | 22.5 | 25 | 43.9 |
| Attracting greater number or mix or caliber of undergraduates into science | 87 | 20.6 | 7 | 12.3 |
| Do not know | 59 | 13.9 | 2 | 3.5 |
| Other | 26 | 6.1 | 7 | 12.3 |
| Not answered | 32 | 7.6 | 1 | 1.8 |
| Total number of surveys | 423 | | 57 | |

## PROPOSED RECOMMENDATIONS

Finally, respondents were asked to list one action that each stakeholder group could take to best facilitate IDR. Responses were categorized and are illustrated below in graphs for institutions, units and departments, funders, journal editors, principal investigators and team leaders, educators, postdoctoral scholars, and students. These were free-response questions; staff analyzed and categorized the responses. Percentages are based on the numbers of responses provided for each category.

The top three recommendations for institutions (n = 341) were to foster a collaborative environment (26.5 percent), to provide faculty incentives (including hiring and tenure policies) that reflect and reward involvement in IDR (18.4 percent), and to provide seed money for IDR projects (11.1 percent). See Figure E-14.

The top three recommendations for departments (n = 294) were to adopt new organizational approaches to IDR (32.1 percent), to recognize and reward faculty and other researchers for interdisciplinary work (20.8 percent), and to adapt or increase departmental resources to support IDR (12.3 percent). See Figure E-15.

The top three recommendations for funding agencies (n = 266) were to provide more support for IDR (39.1 percent), to develop and implement a more effective review process for IDR proposals (17.7 percent), and to rethink funding allocation strategies (11.3 percent). See Figure E-16.

The top two recommendations for journal editors (n = 196) were to adjust the expertise of editorial and review panels and incorporate more reviewers with IDR experience (38.8 percent) and to feature novel innovations and initiatives (36.2 percent); 17.3 percent of respondents reported that they were satisfied with the current situation. See Figure E-17.

The top three recommendations for principal investigators (n = 186) were to increase leadership and team-forming activities (44.1 percent), to develop and clearly state their research goals and their overall vision (34.4 percent), and to build networks with researchers in other disciplines (20.4 percent). See Figure E-18.

Respondents (n = 190) recommended that educators develop curricula that incorporate interdisciplinary concepts (64.7 percent), take part in teacher-development courses on interdisciplinary topics (40 percent), and provide student opportunities in IDR (23.7 percent). See Figure E-19.

Respondents (n = 157) encouraged postdoctoral scholars to get a broad background and learn new skills (14.0 percent), to find a postdoctoral fellowship in a field different from their own graduate work (12.7 percent), and to develop collaborations and seek additional mentors (12.1 percent). See Figure E-20.

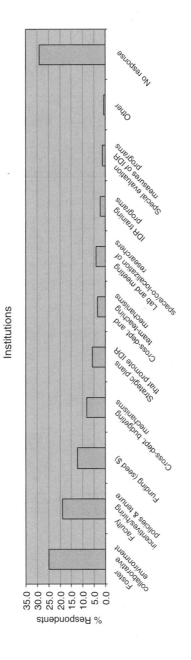

**FIGURE E-14** Institutional recommendations to best facilitate IDR.

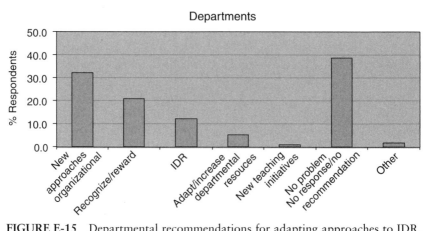

**FIGURE E-15**   Departmental recommendations for adapting approaches to IDR.

Finally, respondents (n = 171) recommended that students cross boundaries between disciplines (25.1 percent), take a broad range of courses (23.4 percent), but also develop a solid background in one discipline (12.3 percent). See Figure E-21.

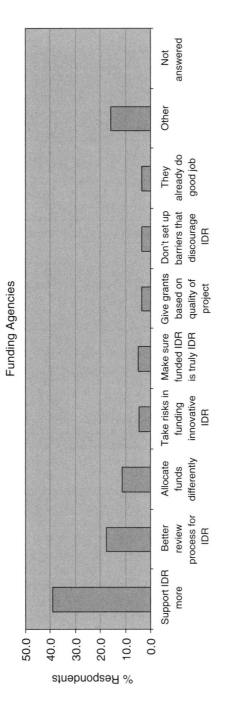

**FIGURE E-16** Recommendations for funding agencies to provide more support to IDR.

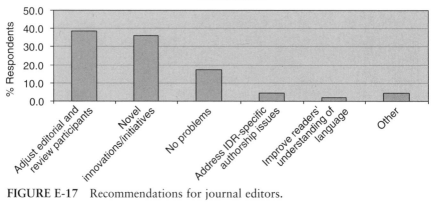

**FIGURE E-17**   Recommendations for journal editors.

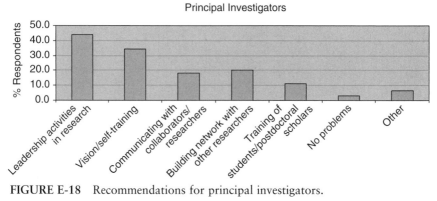

**FIGURE E-18**   Recommendations for principal investigators.

Educators

**FIGURE E-19**   Recommendations for educators.

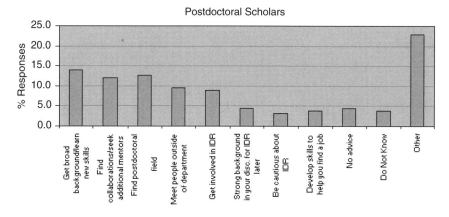

**FIGURE E-20**   Recommendations for postdoctoral scholars.

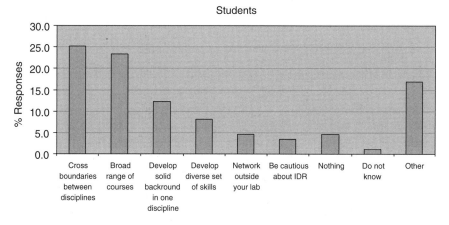

**FIGURE E-21**   Recommendations for students.

## THE "INDIVIDUAL" IDR SURVEY

### Demographics

1. Please tell us about yourself:
   __ Researcher/faculty member
   __ Administrator
   __ Student
   __ Postdoc

   Rank:
   __ Senior
   __ Mid-level
   __ Junior

   Age: _____

   Describe your research:
   __ Primarily interdisciplinary
   __ Partially interdisciplinary
   __ Not interdisciplinary

2. Which of these best describes your institution?
   a. __ Public Academic
   b. __ Private Academic
   c. __ Industrial R&D organization
   d. __ Government R&D organization
   e. __ Independent research institute
   f. __ Other (Please describe): _____

3. What is the size of your institution?
   a. Annual budget:

   | __ $0-1 Million | __ $100-250 M | __ $750 M-1 Billion |
   |---|---|---|
   | __ $1-10 M | __ $250-500 M | __ >$1 B |
   | __$10-100 M | __$500-750 M | __Do Not Know |

   b. If research institution, number of:

| | | | | | | | |
|---|---|---|---|---|---|---|---|
| Faculty/ Researchers | 0 | 1-50 | 50-200 | 200-500 | 500-2000 | Over 2000 | Do Not Know |
| Undergraduates | 0 | 1-500 | 500-2000 | 2000-5000 | 5000-10,000 | Over 10,000 | Do Not Know |
| Graduate Students | 0 | 1-200 | 200-500 | 500-1000 | 1000-2500 | Over 2500 | Do Not Know |
| Postdoctoral Researchers, Fellows, and Trainees | 0 | 1-10 | 11-50 | 51-100 | 101-500 | Over 500 | Do Not Know |

## Relationship to Interdisciplinary Research

4.  How are you involved with interdisciplinary research?

5.  Based on your personal experiences, rate your present institution and prior institutions (that you feel able to judge) on general supportiveness of interdisciplinary research (IDR) using a scale from 0 (IDR-hostile) to 10 (IDR-friendly):

    Current institution
        name: _____
        rating:  0   1   2   3   4   5   6   7   8   9   10
             (hostile)                    (very supportive)
    Previous institution
        name: _____
        rating:  0   1   2   3   4   5   6   7   8   9   10
             (hostile)                    (very supportive)
    Previous institution
        name: _____
        rating:  0   1   2   3   4   5   6   7   8   9   10
             (hostile)                    (very supportive)

## Interdisciplinary Research at Your Institution

6.  Are there impediments to interdisciplinary research at your institution?

    Yes _____ No _____     Do Not Know _____
    If yes, please indicate the top 5 impediments in order of importance.
    __ Budget control
    __ Indirect cost recovery distribution
    __ Publication in disciplinary/interdisciplinary journals
    __ Compatibility with college/dept strategic plans
    __ Promotion and tenure criteria
    __ Credit for joint authorship
    __ Unit reporting relationships
    __ Space
    __ Honoring award agreements
    __ Restrictions on faculty autonomy
    __ Other_____

7.  Does your institution provide seed money to help start up interdisciplinary programs? If yes, please briefly describe the amounts available and major criteria employed in making awards.
    Yes _____ No _____     Do Not Know _____

If yes, please indicate:
Amount:
Duration:
Award Criteria:

8. Does your institution make joint appointments for interdisciplinary faculty/staff members in which salary support is shared between departments, units, and/or schools?
Yes _____   No _____      Do Not Know _____
If yes, what proportion of the faculty/staff have such joint appointments?
__0-10%        __10-25%        __Over 25%

9. Using the table below, please list and describe up to three interdisciplinary program(s) at your institution with which you are currently involved. These programs could be centers, organized research units (ORUs), teaching programs, etc.

|  | A | B | C |
|---|---|---|---|
| Program/Center Name: | | | |
| URL: | | | |
| Contact person: | | | |
| Phone #/e-mail: | | | |
| i. Number of involved depts/ schools/colleges | __ 1 __Don't know<br>__ 2-4<br>__ 5-10<br>__ Over 10 | __ 1 __Don't know<br>__ 2-4<br>__ 5-10<br>__ Over 10 | __ 1 __Don't know<br>__ 2-4<br>__ 5-10<br>__ Over 10 |
| ii. List the primary depts. involved | | | |
| iii. Extra-institutional groups involved? | __ Yes<br>__ No<br>__ Don't know | __ Yes<br>__ No<br>__ Don't know | __ Yes<br>__ No<br>__ Don't know |
| iv. Number of Researchers | __1-5<br>__5-10<br>__10-20<br>__Over 20<br>__Don't know | __1-5<br>__5-10<br>__10-20<br>__Over 20<br>__Don't know | __1-5<br>__5-10<br>__10-20<br>__Over 20<br>__Don't know |

|  | A | B | C |
|---|---|---|---|
| v. Faculty Lines? | __Yes<br>__No<br>__Don't know<br>__Not applicable | __Yes<br>__No<br>__Don't know<br>__Not applicable | __Yes<br>__No<br>__Don't know<br>__Not applicable |
| vi. Source of<br>Funding? | __ DoD<br>__ DoE<br>__ NASA<br>__ NIH<br>__ NSF<br>__ Foundation:<br>__ Institutional:<br>__ Don't know<br>__ Other: | __ DoD<br>__ DoE<br>__ NASA<br>__ NIH<br>__ NSF<br>__ Foundation:<br>__ Institutional:<br>__ Don't know<br>__ Other: | __ DoD<br>__ DoE<br>__ NASA<br>__ NIH<br>__ NSF<br>__ Foundation:<br>__ Institutional:<br>__ Don't know<br>__ Other: |
| vii. Central Facility? | __ Yes<br>__ No<br>__ Don't know | __ Yes<br>__ No<br>__ Don't know | __ Yes<br>__ No<br>__ Don't know |
| viii. Space<br>Allocation | __ Project-driven<br>__ Researcher-<br>specific<br>__ Don't know | __ Project-driven<br>__ Researcher-<br>specific<br>__ Don't know | __ Project-driven<br>__ Researcher-<br>specific<br>__ Don't know |
| ix. Training Slots? | __ Yes<br>__ No<br>__ Don't know | __ Yes<br>__ No<br>__ Don't know | __ Yes<br>__ No<br>__ Don't know |

## Evaluation of Interdisciplinary Research Programs

10. What are the dominant methods of evaluation employed by your institution to evaluate interdisciplinary programs? (check all that apply)

__ Visiting Committee
__ Internal Committee
__ Benchmarking Surveys
__ Interviews
__ Informal Feedback
__ Principal Investigator Assessment
__ Do not know
__ Other (Please describe):

11. What are the dominant methods you use to evaluate the success of interdisciplinary programs? (select up to three or add your own).

___ Level of (or potential for) scientific discovery or innovation

___ Quality of leadership

___ Attracting a greater number/mix/caliber of undergraduates into science

___ Enhancing the richness of the undergraduate/graduate experience

___ Increasing the ability to attract outstanding faculty/postdocs

___ Societal relevance of problem being addressed

___ Enhancing institution's reputation

___ Increasing institution's research funding levels

___ Do not know

___ Other (Please describe):

## Proposed Recommendations

12. If you could recommend one action each of the following could take that would best facilitate interdisciplinary research, what action would that be?

a)  Institutions:

b)  Units/Departments:

c)  Funding Agencies:

d)  Journal Editors:

e)  Principal Investigators/Team Leaders:

f)  Educators:

g)  Postdocs:

h)  Students:

# F

# Committee Interviews
with Administrators, Scholars, and
Center Directors

Over the course of the study, staff supplemented available scholarship with interviews to gain information on the history of interdisciplinary research (IDR) and related scholarship. A primary goal was to collect information on policies, procedures, and effective practices for education programs, research management, and evaluation. Interviewees' names are listed in the order in which they were reached. In most cases, interviews were conducted by teleconference. The symbol * indicates an e-mail interview; the symbol # indicates an in-person meeting.

## IDR HISTORY AND SCHOLARSHIP

Scholars and historians were queried for information on available literature resources and quantitative studies. There is a rich qualitative and philosophical literature on interdisciplinarity,[1] but quantitative studies are few. Much of the research on structural models of interdisciplinarity is based on case studies published in the late 1970s and early 1980s. It was during that period that the National Science Foundation, through its Office

---

[1]For a review of the literature see for example: Klein, J. T. *Interdisciplinarity: History, Theory, and Practice.* Detroit: Wayne State University Press. 1990; Lattuca, L. *Creativity Interdisciplinarity: Interdisciplinary Research and Teaching Among College and University Faculty.* Nashville: Vanderbilt University Press. 2001. Klein, J. T. "Prospectus for Transdisciplinarity." *Futures* 2004, 36:515-526; Rhoten, D. 2004. "Interdisciplinary Research: Trend or Transition." *Items and Issues* 5(1-2):6-11.

of Interdisciplinary Research, provided funding for international meetings on the organization of IDR.[2] Most quantitative research to date has examined interdisciplinarity by using citation-database analysis.[3]
We contacted

- *Margaret Somerville, Samuel Gale Professor of Law and Professor of Medicine, McGill Center for Medicine, Ethics, and Law, McGill University
- *Julie Thompson Klein, professor of humanities, Wayne State University

## IDR PROGRAMS AND CENTERS

IDR program and center directors were asked to discuss their experience in IDR, evaluating prospective researchers, accessing funding, facilitating IDR, determining research goals and duration, evaluating the success of the research team, and publishing research results. We also asked for examples of models and effective practices.

From those discussions, a few themes emerged: leadership, institutional support, and departmental buy-in. To create a successful academic interdisciplinary center or program required a visionary leader. In addition to being persistent and persuasive, the leader had to have sufficient stature in the institution and in a research field and the support of the university president or provost. The leader had to coordinate her/his vision with relevant institutional departments; in effect, the leader needed to develop partnerships and sell participation in the program or center. The leader had to successfully negotiate shared costs, faculty hires, space allocation, and funding. Finally, the leader had to recruit and sustain faculty and student participation.

---

[2]See *Managing High Technology: An Interdisciplinary Perspective.* Eds. Mar, B.W., Newell, W.T., and Saxberg, B.O. Elsevier: New York. 1985. This volume is based on papers from the Third International Conference on Interdisciplinary Research, Seattle, Washington, U.S.A., 1-3 August, 1984.
[3]Baumann, H. 2003. Publish and Perish? The impact of citation indexing on the development of new fields of environmental research. *Journal of Industrial Ecology* 6, 3-4:13-26; Chubin, D. E., Porter, A. L., and Rossini, F. A. 1984. "Citation Classics" Analysis: An Approach to Characterizing Interdisciplinary Research. *Journal of the American Society for Information Science* 35, 6:360-368; McCain, K. W. and Whitney, P. J. 1994. Contrasting Assessments of Interdisciplinarity in Emerging Specialties: The Case of Neural Networks Research. *Knowledge: Creation, Diffusion, Utilization* 15, 3:285-306; Steele, T. W. and Stier, J. C. 2000. The Impact of Interdisciplinary Research in the Environmental Sciences: A Forestry Case Study. *Journal of the American Society for Information Science* 51, 5:476-84.

We contacted

- James McClelland, director, Center for the Neural Basis of Cognition, Carnegie Mellon University, *http://www.cnbc.cmu.edu/*
- Frances Leslie, director, Transdisciplinary Tobacco Use Research Center, University of California, Irvine, *http://www.tturc.uci.edu/*
- Jim LeBaugh, Water Resources Division, United States Geological Survey, and participant, Shingobee Headwaters Aquatic Ecosystems Project (SHAEP), *http://wwwbrr.cr.usgs.gov/projects/IRI/*
- C. Channa Reddy, director, Huck Institute for Life Sciences, Pennsylvania State University, *http://www.lsc.psu.edu/*
- Michael Merson, director, Center for International Research on AIDS, Yale University, *http://cira.med.yale.edu*
- *John Ballato, director, Center for Optical Materials and Science and Engineering Technologies (COMSET), Clemson University, Carolinas Optics Center, *http://www.ces.clemson.edu/comset/*

## INTERDISCIPLINARY EDUCATION PROGRAMS

Education-program directors were asked to provide an overview of their interdisciplinary programs, the impetus for establishing them, their goals and duration, evaluation criteria for the competition, and information on where the programs have been implemented. We asked how an education program encouraged the development of interdisciplinary curricula and pedagogic tools. Finally, we asked for examples of models and effective practices and for suggestions of additional contacts.

Among the themes that emerged were a concern that science and engineering programs were in general not appealing to undergraduates and that undergraduate and graduate programs do not sufficiently prepare students for careers in industry. Interdisciplinary problem-based approaches to learning were seen as a way to encourage more students to take science classes and to prepare students for a variety of careers.

We contacted

- Gerry Wheeler, executive director, National Science Teachers Association, Re: ExploraVision, *http://www.exploravision.org/*
- #Wyn Jennings, director, IGERT Program, Division of Graduate Education, National Science Foundation, *http://www.nsf.gov/home/crss prgm/igert/start.htm*
- Ed Miles, former chair, Task Force on Environmental Education, and professor, School of Marine Affairs, Graduate School of Public Affairs, University of Washington, *http://depts.washington.edu/poeweb/*

## IDR EVALUATION

Evaluation scholars were asked to share IDR evaluation tools and case studies with the committee relevant to IDR and program evaluation, promotion and tenure, budget models, and education and career development. In most cases, IDR evaluation tools were in development and unavailable. Social-networks analysis was often cited as an evaluation concept that had been used successfully to evaluate IDR. But scholars were quick to point out that this analysis, while providing a measure of interconnectedness and interdisciplinarity, did not necessarily measure research quality and impact. There was agreement that more work was needed to develop specific criteria and measures for IDR.

We contacted

- #Irwin Feller, professor emeritus of economics, Pennsylvania State University
- Diana Rhoten, Helen Doyle, and Denise Caruso, Hybrid Vigor Institute, *http://www.hybridvigor.org/*
- Barry Bozeman, Regents' Professor of Public Policy, Georgia Institute of Technology
- Ed Hackett, professor, Department of Sociology, Arizona State University
- Marye Anne Carroll, director, Biosphere-Atmosphere Research and Training (BART), University of Michigan; and Kristin Kusmierek, BART IGERT program evaluator, *http://www.bart-wmich.org/*

## IDR POLICIES

Policy makers and research administrators were asked for information on policies and procedures to facilitate IDR. We asked those at academic institutions to discuss their experience in promoting interdisciplinary initiatives. We asked research administrators to share their experiences and policies for evaluating prospective interdisciplinary researchers, accessing funding to support interdisciplinary projects, hiring interdisciplinary faculty, and facilitating IDR. For example, we asked whether faculty teaching time was shared between departments, how space for projects involving faculty from multiple departments was allocated, and whether faculty hires were made collaboratively between departments. We also inquired about how research project and program goals and duration were determined. For example, we asked how the success of interdisciplinary projects was evaluated and whether publication of research results was a key component in that evaluation. Finally, we asked for examples of models and best practices.

In this category, a general theme was flexibility. Specific solutions need to be tailor-made to fit institutional context, but there are examples of effective policies and organizational structures. Administrators suggested meshing vertical departmental structures with horizontal, cross-cutting programmatic themes. A discretionary fund, or seed money, controlled by the provost was critical for promoting and supporting cross-cutting initiatives. Specific guidelines for promotion and tenure that accounted for interdisciplinary scholarship had been enacted. Team teaching was encouraged, and in many cases credit hours were counted by all the involved departments. Still, administrators concurred that more needed to be done to provide for cost-sharing between departments and between institutions, especially at the grant level. Some concern was expressed about national evaluation of IDR programs and centers, many of which exist outside standard institutional structures.

We contacted

• National Science and Technology Council Subcommittee on Research Business Models, Committee on Science, Office of Science and Technology Policy
• June Howard, associate dean for interdisciplinary initiatives, University of Michigan
• Cornelius Sullivan, vice provost for research, University of Southern California
• Maria Pallavicini, professor and dean, School of Natural Sciences, University of California, Merced

## IDR IN INDUSTRY AND NATIONAL LABORATORIES

Directors and researchers were asked about the importance of IDR in industry and national labs. Specifically, what actions were taken to facilitate IDR? How were people organized to work together on IDR problems? What are examples of where IDR worked and where it did not work? Has the role of IDR teams evolved? And finally, what lessons can national labs provide to academia as to how to best facilitate IDR? The results of these interviews are summarized and presented in Chapter 3.

We contacted:

• #John Armstrong, vice president, Science and Technology, IBM (ret.)
• *Norm Burkhard, division leader, Energy and Environment Directorate, Lawrence Livermore National Laboratory
• *Michael Crow, president, Arizona State University

- Bernard S. Meyerson, IBM Fellow, VP and Chief Technologist, IBM Systems and Technology Group
- Edward C. Stone, David Morrisroe Professor of Physics; Vice Provost for Special Projects; former Vice President and Director of the Jet Propulsion Laboratory (1991-2001).
- *Tom Wilbanks, chair, Corporate Fellows Council, Oak Ridge National Laboratory

# G

# Focus Groups on Facilitating Interdisciplinary Research

NATIONAL ACADEMIES
KECK FUTURES INITIATIVE CONFERENCE

Signals, Decision, and Meaning in Biology, Chemistry,
Physics, and Engineering

Irvine, California
November 15, 2003

The National Academies Keck *Futures Initiative* Conference brings together over 100 of the nation's best and brightest researchers from academic, industrial, and government laboratories to ask questions about—and to discover interdisciplinary connections between—important areas of cutting-edge research.

At the first Keck *Futures* meeting in November 2003, the Committee on Facilitating Interdisciplinary Research hosted three focus groups to brainstorm policies and practices that funding organizations, educators, academic administrators, researchers, and students could implement to overcome barriers to interdisciplinary research (IDR). The focus was on the role that policies and practices related to training, education, evaluation, team-building, funding, hiring, and employment could play in facilitating IDR. The committee was especially interested in learning about effective programs and policies; to this end, the moderator was encouraged to steer discussion from that of barriers to one of suggestions and solutions. The data gathered from the focus groups were used to help the committee to develop findings and recommendations. It is important to keep in mind in reviewing these comments that this group is made up entirely of those interested in IDR.

## FOCUS-GROUP QUESTIONS

The following discussion questions were provided to each moderator for discussion.

### Training and Education

Should undergraduate students be encouraged to pursue an interdisciplinary degree? What policies can institutions adopt that would facilitate team teaching, curricular development, and cross-departmental course offerings? What programs and/or policies would be most effective at facilitating interdisciplinary training of graduate students and postdoctoral scholars?

### Hiring and Employment

What can institutions do to facilitate hiring and review of interdisciplinary faculty? Are joint appointments a good idea? Are multi-departmental review panels effective? Should outside experts be appointed to review panels for interdisciplinary tenure candidates? What strategies can an interdisciplinary tenure-track researcher employ to enhance the review process? What can faculty and departments do to enhance the process?

### Evaluation

What are effective criteria for evaluating interdisciplinary papers? Interdisciplinary researchers? Interdisciplinary programs? What can investigators, institutions, and funding agencies do to enhance the review/evaluation procedure? Does interdisciplinary research require different or additional criteria for evaluation than disciplinary research?

### Establishing a Team

What programs and policies can institutions and funding agencies adopt to facilitate collaboration between disciplines? Are seed grants effective? Are meetings effective? What are the critical aspects of team formation?

## Funding

What are the most effective funding strategies for facilitating IDR? Should funders focus on research grants in emerging areas, seed grants for teams, infrastructure development, training and education, and/or internal polices and procedures to facilitate submission and review of interdisciplinary proposals (e.g., panel review, site visits, etc.)? Are there polices that federal agencies or institutions could adopt that would facilitate IDR, such as budgeting structures, cost-sharing, allowing for co-PIs, etc.?

## FOCUS-GROUP FORMAT

The moderators were Bruce Alberts, Bill Wulf, and Harvey Fineberg, presidents of the National Academy of Sciences, the National Academy of Engineering, and the Institute of Medicine, respectively. Each focus group consisted of about 20 researchers in many fields. The results of those discussions follow this summary. Each used a different discussion technique, and the results reflect that.

## SUMMARY OF FOCUS GROUP A DISCUSSION
### (MODERATOR, HARVEY FINEBERG, PRESIDENT, INSTITUTE OF MEDICINE)

### Definition

*Scale and scope:* A researcher in biomolecular systems at Pacific Northwest National Laboratory suggested that "what drives IDR is the scale and scope of the problem." He said that a truly interdisciplinary problem would require five to 10 investigators; for biologic systems, it would involve not only biologists, but also mathematicians, instrument builders, and others. "As long as funding agencies say they want to solve large problems, we'll see communities come together" to do so.

### Models

*Promoting collaboration:* A professor of chemistry and neurosciences at the University of Illinois Beckman Institute noted that "it's a physical space, with no funding, and people from many labs." They share space and equipment. Each participant does not do disciplinary work but builds on it and is encouraged to ask for collaboration. The Institute "formalizes the idea that you'll work with someone else." Faculty can apply for a semester's training in another discipline, as long as their faculty head signs off on it.

*Beyond departments:* In 2005, when the University of California, Merced, opens its campus, its School of Natural Sciences will have no departments and will integrate science and engineering. The dean of the new school said that it had made a commitment to hire faculty with excellence in a particular discipline to avoid the "risk of being shallow across whatever you do". One challenge she noted was that the 15 faculty who had already been recruited tended to interview faculty applicants according to criteria of excellence that differed between disciplines. "Until we get that it will be hard to be successful, especially at the junior level."

*Crossing theoretical disciplines:* A scientist at the Salk Institute praised the Sloan Foundation's program in theoretical neurobiology, which brings young theoreticians from the physical, mathematical, and computer sciences into neurobiology at five university-based research centers. Some have gone on to start their own laboratories he said, although faculties sometimes blocked cross-disciplinary hiring recommended by "visionary" search committees.

## Policies and Procedures

*Tenure as an obstacle:* A professor in the Massachusetts Institute of Technology Program in Science, Technology, and Society said that the largest obstacle to IDR in universities has been tenure. When one is a postdoctoral scholar or an assistant professor, she said, it is risky to work outside one's own department. She applauded the initiative of the Harvard Medical School in founding its new Department of Systems Biology, which is inherently interdisciplinary.

*Beyond departments:* A professor in the Harvard Medical School Department of Biological Chemistry and Molecular Pharmacology said, "I think universities could get rid of departments." She admitted that her view came out of her work in cancer research, which is highly interdisciplinary.

*Three effective procedures:* A professor in the Harvard Department of Physics and Applied Physics recommended three procedures that he had found effective in promoting interdisciplinary work:

- 24-hour retreats on campus for groups of faculty. He described a successful retreat on neurosciences, in which faculty established personal connections and talked about long-term interests in ways that they could not easily do in the midst of busy schedules.
- Working in other departments and experiencing related or relevant fields.
- Getting some seed money from the university (for example, the dean's fund) for a postdoctoral or graduate student who would like to work in different fields.

*Teaching in mixed groups:* A biology professor who works in biomedical research in a Canadian Organized Research Unit noted the risk of teaching biology to computer-science students because of the difficulty of communication. He had found that a computer scientist might say he was going to do one thing and a biologist something else, and it could turn out that they intended to do the same thing. For a biologist, however, the risk was necessary to model biologic systems. "Certainly, computer-science students are fascinated by questions in biology. You need to take that step and go out and teach in mixed groups and learn their language."

*Policies at state universities:* The Texas A&M Department of Chemical Engineering and Chemistry had found it possible to share National Institute of Health (NIH) grants "so that everyone gets something." When the university budgets were cut by the state, however, principal investigators (PIs) had to show the revenue generated by their own research to maintain their share of state funding, and the sharing mechanism was in jeopardy.

*Promoting communication:* A Salk Institute investigator saw communication as a key, especially better communication between funding committee members of different backgrounds and better communication of the intellectual content of one's own work.

## Training and Education

*Following one's curiosity:* Entelos,[1] a private firm working in computer modeling of diseases, needs both mathematicians and engineers for its interdisciplinary work. It prefers to hire "a great person rather than someone who's already been trained in two disciplines." The chief scientist referred to her own experience as a graduate student, when she and her colleagues first attained a solid grounding in their field and then benefited by following their curiosity to work on problems in other departments.

*"Excellence at the interface":* In training young IDR investigators, a member of the Pharmacology Department of the University of Texas argued in favor of "finding individuals who have more than one discipline in one brain, to make that creative stuff happen." People who are excellent in one discipline, he said, may not make good collaborators. The ideal scenario is to "create that depth in individuals at the interface. Students brought up in that ethic and studying at the interface learn how to be good in more than one thing."

---

[1]Entelos is a firm that develops large-scale computer models of human disease using a patented PhysioLab technology. In partnership with biotechnology and pharmaceutical companies, it seeks to speed development of new treatments for such diseases as asthma, obesity, and diabetes.

*Planning IDR from scratch:* In planning of curricula for the new School of Natural Sciences of the University of California, Merced, two challenges had arisen: (1) faculty did not want to "give up any content in their courses" and (2) planning an interdisciplinary undergraduate curriculum turned out to be harder than planning an interdisciplinary graduate curriculum because faculty felt a need to cover the basics first.

*Inviting students to initiate research:* The norm is for a PI to recruit graduate students on the basis of a project whose goals have been determined. In that system, students often feel like hired hands working for someone else. A Harvard professor argued for the reverse: Challenge graduate students, who may be more up to date than their professors, to design their own research projects and win the support of PIs and laboratories.

*Flinging graduate students through the "gates of Hell":* A professor at Thomas Jefferson University argued in favor of exposing engineers to biologic problems by putting them through biology courses at the same level as medical students. They would collect their own data and gain a realistic view of gathering data. "Then they become the 'glue people' that you need in multidisciplinary groups." Several people voiced strong agreement with that proposal.

## SUMMARY OF FOCUS GROUP B DISCUSSION

### (MODERATOR, BILL WULF, PRESIDENT, NATIONAL ACADEMY OF ENGINEERING)

The focus group started with the premise that there are six targets for which specific short- and long-range goals could be set to foster IDR: education, culture, hiring and employment, publication, evaluation, and funding. The goals are listed below by target.

Some of the proposed goals are seemingly straightforward and could be implemented at the individual, department, or institution level with little financial or logistical difficulty. For example, cross listing graduate-school classes across departments or writing abstracts for a more general scientific audience would take little effort and may reap large rewards. Others would require more long-range strategic planning, such as adjusting the National Science Foundation (NSF) Research Experience for Undergraduates (REU) program to contain more mentors so that the undergraduate students would have a broader exposure to cross-discipline projects.

### Education

- Cross-list all graduate-school courses in all departments.
- Allow greater freedom with respect to electives in graduate school.

- Offer more classes that have no prerequisites.
- Foster joint-degree programs.
- Create a buddy system—for example, with a graduate student in biology matched with one in mathematics. Promote informal lunch meetings between them. Make them explain their work to each other.
- The summer NSF REU program should require two mentors, not just one.
- Mandate industrial internships before granting the PhD. Industry is intrinsically more interdisciplinary than academe.

### Culture

- The only thing that will really foster change is years of lunches shared by disparate groups.
- It is not the faculty who are in the best position to spur IDR, but rather the "lab rat" who is actually doing experiments.
- Mimic the 1993 "Grand Challenge" by having a central entity define long-term unresolved problems and issue them as challenges.
- Co-locate departments; don't allow physical space to constitute a barrier between departments (for example, biochemistry on one floor, microbiology on the next, comparative biology on the next).
- Establish postdoctoral salary parity across fields (physical-science postdoctoral scholars are paid much more than biology postdoctoral scholars).
- Encourage graduate students to switch departments when doing postdoctoral work.

### Hiring and Employment

- Create incentives for departments to create and fill interdisciplinary positions (along the lines of affirmative action).
- Highlight the availability of people with interdisciplinary skills (such as people who run core facilities).

### Publication

- Do not promote new journals that are classified as being in single disciplines. Submit papers only to interdisciplinary journals.
- Promote and fund databases that cover multiple journals in many fields. (For example, the National Library of Medicine searches almost no mathematical fields.) We are in an article-based, not a journal-based, publication environment.
- Require that abstracts be written for a more general audience.

Evaluation (for promotion and in peer review)

- Reward at the institutional level.
- Make sure that departments do not hold up promotion of cross-department faculty. (Sometimes an institution has to intervene or simply make promotion and tenure decisions only at the institution level.)
- Document evaluation norms by discipline. For example, in physics, conference proceedings are much more prestigious than publications; in biology it is the opposite.
- Reward people for publishing in a variety of journals, as opposed to only journals with high impact factors—for example, two articles in journals sponsored by very different professional societies (such as the Society for Neuroscience and the American Physical Society).

## Funding

- Promote streamlined procedures for interdisciplinary signoff at universities. Getting a joint grant is too laborious, and the deans want to know only who is subject to the direct costs and overhead.
- Promote a mechanism for 5-6 years of support based solely on the drive to learn another discipline or to learn core new skills not normally attributed to the "home" department.
- Students need to know that some places, such as publishing and industry, financially reward people who have multidisciplinary backgrounds.

## SUMMARY OF FOCUS GROUP C DISCUSSION

### (MODERATOR, BRUCE ALBERTS, PRESIDENT, NATIONAL ACADEMY OF SCIENCES)

The discussion focused primarily on evaluation and funding mechanisms. The following is a compilation of the participants' top recommendations for facilitating IDR.

## Evaluation

- Go beyond research issues in evaluating IDR; education is a key factor as well.
- Focus on the quality of the people who are submitting grant proposals.

## Funding Mechanisms

- The next generation is the key to IDR, so look at the experience of the NIH Alliance for Cellular Signaling in working with junior scholars.

- Effective programs that have a large impact on the potential impact of a beginning researcher to hire people and obtain computers and other necessary equipment need not be high-cost. For example, the NIH FIRST (R29) award provides a research support for newly independent, biomedical and behavioral science investigators to initiate their own research and demonstrate the merit of their own research ideas.[2]
- Focus more on middle-level people who have tenure, because they are able to take the risks entailed in IDR.
- Focus funding on fellows and on travel grants that provide them with the necessary independence.
- Create independent IDR institutions where people can come together on equal footing.

## Institutional Mechanisms

- Focus attention on institutional roles—the leadership of an organization is critical.
- Create universitywide interdisciplinary research positions.

## Other issues:

- When disciplines come together, they need to do so on an equal basis
- Treat postdoctoral fellows as the glue between researchers who should be joint advisers.
- Study history.
- Make medical schools more hospitable to IDR.

---

[2]Guidelines for FIRST awards Web page *http://grants2.nih.gov/grants/policy/r29.htm.*

# H

# Bibliography

1. "Advanced Photon Source Home page." Web page, [accessed 6 May 2004]. Available at http://epics.aps.anl.gov/aps.php.
2. "Association for Integrative Studies (AIS)." Web page, [accessed 8 August 2003]. Available at http://www.units.muohio.edu/aisorg/intro.html.
3. "Budgeting with the UB Model at the University of Michigan." Web page, [accessed 24 July 2003]. Available at http://www.umich.edu/~provost/budgeting/ubmodel.html.
4. "Department of Defense Strategic Plan." Web page, [accessed 22 July 2003]. Available at http://www.conginst.org/resultsact/PLANS/plans_dod.html.
5. "Department of Energy Strategic Plan." Web page, [accessed 15 July 2003]. Available at http://www.cfo.doe.gov/stratmgt/plan/DOEsplan.htm.
6. "Environmental Protection Agency/OCFO—Strategic Plan." Web page, [accessed 15 July 2003]. Available at http://www.epa.gov/ocfo/plan/plan.htm.
7. "Fund for the Advancement of the Discipline." Web page, [accessed 23 February 2004]. Available at http://www.asanet.org/members/fad.html.
8. "Good Work Project." Web page, [accessed 18 July 2003]. Available at http://www.goodworkproject.org.
9. "History of X-ray Crystallography and Associated Topics." Web page, [accessed 8 March 2004]. Available at http://www.dl.ac.uk/SRS/PX/history/history.html.
10. "Human Frontier Science Program." Web page, [accessed 30 July 2003]. Available at http://www.hfsp.org/home.php.
11. "Institute for Operations Research and the Management Sciences (INFORMS)." Web page, [accessed 8 August 2003]. Available at http://www.informs.org/.
12. "Integrative Graduate Education and Research Traineeship (IGERT)." Web page, [accessed 10 March 2004]. Available at http://www.nsf.gov/pubs/2004/nsf04550/nsf04550.pdf.
13. "Keck Futures Conferences." Web page, [accessed 3 March 2004]. Available at http://www7.nationalacademies.org/keck/Keck_Futures_Conferences.html.

14. "Marine Biological Laboratory Home page." Web page, [accessed 6 May 2004]. Available at http://www.mbl.edu/research/summer/index.html.

15. "Mathematical Sciences Research Institute." Web page, [accessed 6 May 2004]. Available at http://www.msri.org/.

16. "Microsoft Research Home page." Web page, [accessed 6 May 2004]. Available at http://research.microsoft.com/aboutmsr/jobs/internships/.

17. "National Academies Keck Futures Initiative." Web page, [accessed 3 March 2004]. Available at http://www7.nationalacademies.org/keck/index.html.

18. "National Institute of Standards and Technology 2010 Strategic Plan." Web page, [accessed 16 July 2003]. Available at http://www.nist.gov/director/.

19. "National Science Foundation GPRA Strategic Plan FY 2001—2006." Web page, [accessed 15 July 2003]. Available at http://www.nsf.gov/pubs/2001/nsf0104/start.htm.

20. "The Shingobee Headwaters Aquatic Ecosystems Project (SHAEP) Home page." Web page, [accessed 6 May 2004]. Available at http://wwwbrr.cr.usgs.gov/projects/SHAEP/index.html

21. "Society for Social Studies of Science (4S)." Web page, [accessed 8 August 2003]. Available at http://www.lsu.edu/ssss/.

22. "Training for a New Interdisciplinary Research Workforce." Web page, [accessed 1 June 2004]. Available at http://grants1.nih.gov/grants/guide/rfa-files/RFA-RM-04-015.html.

23. "University of Minnesota Budget and Interdisciplinary Initiatives." Web page, [accessed 5 August 2003]. Available at http://www.evpp.umn.edu/evpp/init.htm.

24. "USGS Home page; Cottonwood Lake Study Area." Web page, [accessed 6 May 2004]. Available at http://www.npwrc.usgs.gov/clsa/index.htm.

25. "Vanderbilt Institute of Chemical Biology." Web page, [accessed 20 January 2004]. Available at http://www.vanderbilt.edu/vicb/home.html.

26. "What is Biogeochemistry?" Web page, [accessed 20 January 2004]. Available at http://calspace.ucsd.edu/virtualmuseum/climatechange1/05_1.shtml.

27. Advisory Council for Science and Technology Policy. 2003. *1 + 1 > 2 Promoting Interdisciplinary Research*, Report 54 of Series Advisory Reports. http://www.awt.nl/en/index.html.

28. Balsiger, P. W. 2004. Supradisciplinary research practices: history, objectives and rationale. *Futures* 36, no. 4:407-21.

29. Barth, R. T. and R. Steck. 1979. *Interdisciplinary Research Groups: Their Management and Organization*. Vancouver: International Group on Interdisciplinary Programs.

30. Baumann, H. 2003. Publish and Perish? The impact of citation indexing on the development of new fields of environmental research. *Journal of Industrial Ecology* 6, no. 3-4:13-26.

31. Bayer, P. S. and B. Steinheider. 2004. Working in interdisciplinary research teams: Why heterogeneity matters. Society for Industrial and Organizational Psychology.

32. Bechtel, W. 1986. *Integrating Scientific Disciplines: The Nature of Scientific Integration*. Dordrecht: Martinus Nijhoof.

33. Bell, N. 2003. Mapping Academic Disciplines to a Multi-Disciplinary World. *Proceedings of an NSF/CPST/Professional Societies Workshop* Washington, D.C.: Commission on Professionals in Science and Technology.

34. Birnbaum-More, P. H., F. Rossini, and D. Baldwin. 1990. *International Research Management*. New York: Oxford University Press.

35. Birnbaum, P. H. 1983. Predictors of Long-Term Research Performance. *Managing Interdisciplinary Research*. S. R. Epton, R. L. Payne, and A. W. Pearson, 47-59. New York: John Wiley and Sons.

36. Birnbaum, P.H. 1979. A Theory of Academic Interdisciplinary Research Performance: A Contingency and Path Analysis Approach. *Management Science* 25, no. 3:231-42.
37. Bourke, P. and L. Butler. 1998. Institutions and the map of science: Matching university departments and fields of research. *Research Policy* 26:711-18.
38. Bozeman, B. and E. Corley. 2004. Scientists' Collaboration Strategies: Implications for Scientific and Technical Human Capital. *Research Policy* 33, no. 4:599-616.
39. Brainard, Jeffrey. 2002. U.S. Agencies Look to Interdisciplinary Science. *The Chronicle of Higher Education*: A20.
40. Brandt, E.N. 1997. Research Administration in a Time of Change. *SRA Journal* 29, no. 1,2:233-36.
41. Bruce, A., C. Lyall, J. Tait, and R. Williams. 2004. Interdisciplinary integration in Europe: the case of the Fifth Framework programme. *Futures* 36, no. 4:457-70.
42. Bruhn, J.G. 2000. Interdisciplinary Research: A Philosophy, Art Form, Artifact or Antidote? *Integrative Physiological and Behavioral Science* 35, no. 1:58-66.
43. Butler, D. 1998. Interdisciplinary research 'being stifled'. *Nature* 396, no. 6708:202.
44. Butler, D. 2000. Pasteur Institute to Abandon Departmental Structure? *Nature* 405, no. 6790:990.
45. Cairns, R. B., E. J. Costello, and Elder, G. H. 1996. The Making of Developmental Science. *Developmental Science*. R.B. Cairns, G.H. Elder, and E.J. Costello, 223-34. New York: Cambridge University Press.
46. Campbell, W. H., W. K. Anderson, and G. J. Burckart. 2003. "Institutional and faculty roles and responsibilities in the emerging environment of university-wide interdisciplinary research structures." *Report of the 2001-2002 Research and Graduate Affairs Committee*, American Association of Colleges of Pharmacy.
47. Caruso, D. and D. Rhoten. 2001. Lead, follow, get out of the way: sidestepping the barriers to effective practice on interdisciplinarity. *The Hybrid Vigor Institute* http://www.hybridvigor.net/interdis/pubs/hv_pub_interdis-2001.04.30.pdf.
48. Chase, A., J. Giancola, L. Horst, R. Jacob, A. Martinez, M. A. Millsap, W. C. Smith, S. Goldsmith, D. Haviland, and N. Tushnet. 2003. *IGERT Annual Cross-Site Report: 1998 and 1999 Cohorts*, Abt Associates Inc., Washington D.C.
49. Chase, A., J. Giancola, L. Litin, A. Martinez, and C. Weiland. 2003. *IGERT Implementation and Early Outcomes: 2002*, Abt Associates Inc., Washington D.C.
50. Chubin, D. E. 1986. "Interdisciplinary Analysis and Research: Theory and Practice of Problem-Focused Research and Development." *Interdisciplinary Analysis and Research: Theory and Practice of Problem-Focused Research and Development*. Mt. Airy: Lomond Publications.
51. Chubin, D. E., A. L. Porter, and F. A. Rossini. 1984. "Citation Classics" Analysis: An Approach to Characterizing Interdisciplinary Research. *Journal of the American Society for Information Science* 35, no. 6:360-368.
52. Chubin, D. E., A. L. Porter, F. A. Rossini, and T. Connolly. 1986. *Interdisciplinary Analysis and Research: Theory and Practice of Problem-Focused Research and Development*. Mt. Airy: Lomond Publications.
53. Cohen, L., M. J. Aboelata, T. Gantz, and J. Van Wert. 2003. "Collaboration Math: Enhancing the Effectiveness of Multidisciplinary Collaboration: Applying Collaboration Math to the U.C. Berkeley Traffic Safety Center." Prevention Institute, www.preventioninstitute.org.
54. Collins, J. P. 2002. May You Live in Interesting Times: Using Multidisciplinary and Interdisciplinary Programs To Cope with Change in the Life Sciences. *Bioscience* 52, no. 1:75-83.

55. Colwell, R. 2001. From Microscope to Kaleidoscope: Merging Fields of Vision. *Transdisciplinarity: Joint Probelm Solving among Science, Technology, and Society.* J. Thompson-Klein, W. Grossenbacher-Mansuy, R. Haberli, A. Bill, R.W. Scholz, and M. Welti, 59-66. Basel, Boston, Berlin: Birhauser-Verlag.

56. Committee on Building Bridges in Brain, Behavioral, and Clinical Sciences. 2000. "Bridging Disciplines in the Brain, Behavioral, and Clinical Sciences." National Academy Press, Washington, D.C.

57. Committee on Human Factors. 2003. "Dynamic Social Network Modeling and Analysis: Workshop Summary and Papers." National Academy Press, Washington, D.C.

58. Committee on Large-Scale Science and Cancer Research. 2003. *Large-Scale Biomedical Science: Exploring Strategies for Future Research.* Washington, D.C.: The National Academies Press.

59. Committee on Research in Mathematics, Science, and Technology. 1987. *Interdisciplinary research in mathematics, science, and technology education.* Washington, D.C.: National Academy Press.

60. Committee on Undergraduate Biology Education to Prepare Research Scientists for the 21st Century. 2003. "BIO 2010: Transforming Undergraduate Education for Future Research Biologists." Washington, D.C.: National Academy Press.

61. Committee to Examine the Methodology for the Assessment of Research-Doctorate Programs, Policy and Global Affairs Division. 2003. "Assessing Research-Doctorate Programs: A Methodology Study." Washington, D.C.: The National Academies Press.

62. Couzin, J. 2003. Congress Wants the Twain to Meet. *Science* 301:444.

63. Cummings, J. and S. Kiesler. 2003. *KDI Initiative: Multidisciplinary Scientific Collaborations,* NSF Award No. IIS-9872996. National Science Foundation.

64. Cutler, R. S. 1979. A Policy Perspective on Interdisciplinary Research in U.S. Universities. *Interdisciplinary Research Groups: Their Management and Organization.* R. T Barth, and R. Steck, 295-314. IGRIP.

65. Davis, J. R. 1995. *Interdisciplinary Courses and Team Teaching: New Arrangements for Learning.* Phoenix, AZ: Oryx Press.

66. de Paula, J. C. 2003. Integrating the Sciences at Haverford College. *Research Corporation Report* http://www.rescorp.org/report/rc_report.htm: 3.

67. Defila, R., and A. DiGiulio. 1999. Evaluating Transdisciplinary Research. *Panorama* 1.

68. Derry, S. J. C. D. Schunn, and M. A. Gernsbacher. 2004. *Problems and Promises of Interdisciplinary Collaboration: Perspectives from Cognitive Science.* Mahwah, NJ: Erlbaum.

69. Despres, C., N. Brais, and S. Avellan. 2004. Collaborative planning for retrofitting suburbs: transdisciplinarity and intersubjectivity in action. *Futures* 36, no. 4:471-86.

70. DeWachter, M. 1982. Interdisciplinary Bioethics: But Where Do We Start? A Reflection on Epoche as Method. *Journal of Medicine and Philosophy* 7, no. 3:275-87.

71. Dill, K. A. 1999. Strengthening biomedicine's roots. *Nature* 400:309-10.

72. Dressel, P. 1970. *The Confidence Crisis: An Analysis of University Departments.* San Francisco: Jossey-Bass.

73. Edwards, A. Jr. 1996. *Interdisciplinary Undergraduate Programs: A Directory.* Acton, MA: Copley Publishing Group.

74. Elder, G. H., J. Modell, and R. D. Parke. 1993. Epilogue: An emerging framework for dialogue between history and developmental psychology. *Children in time and place: Developmental and historical insights.* G. H. Elder, J. Modell, and R. D. Parke, 241-49. New York: Cambridge University Press.

75. Elgass, J. R. 1994. Clarke co-directs project at Argonne photon facility. *The University Record* March 28.

76. Epton, S. R., R. L. Payne, and A. W. Pearson. 1983. Contextual Issues in Managing Cross-Disciplinary Research. *Managing High Technology: An Interdisciplinary Perspective*, 209-29. New York: Elsevier.

77. Epton, S. R., R. L. Payne, and A. W. Pearson. 1984. Cross-Disciplinarity and Organizational Forms. Managing Interdisciplinary Research. Chichester: John Wiley and Sons.

78. Epton, S. R., R. L. Payne, and A. W. Pearson. 1983. *Managing Interdisciplinary Research*. Chichester: John Wiley and Sons.

79. European Commission. 2003. "CORDIS FP6: Guidelines on proposal evaluation and selection procedures [FP6]." Web page. Available at http://dbs.cordis.lu/.

80. European Commission. 2003. "CORDIS FP6: Provisions for implementing integrated projects: Background document. " Web page. Available at http://dbs.cordis.lu/.

81. Feder, T. 2000. Strasbourg Interdisciplinary Institute Gets Off the Ground. *Physics Today* April: 58.

82. Feist, G. J. and M. E. Gorman. 1998. The Psychology of Science: Review and Integration of a Nascent Discipline. *Review of General Psychology* 2, no. 1:3-47.

83. Feller, I. 2002. New Organizations, old cultures: stategy and implementation of interdisciplinary programs. *Research Evaluation* 11:109-16.

84. Feller, I. 2002. Performance Measurement Redux. *American Journal of Evaluation* 23:435-52.

85. Feller, I. 2004. Whither Interdisciplinarity (In an Era of Strategic Planning)? *Proceedings of 2004 Annual AAAS Meeting* Washington D.C.: AAAS.

86. Fiscella, J. and S. Kimmel. 1999. "Interdisciplinary Education: A Guide to Resources." *Interdisciplinary Education: A Guide to Resources*, The College Board, New York.

87. Flexner, H. 1979. The Curriculum, the Disciplines, and Interdisciplinarity in Higher Education: Historical Perspective. *Interdisciplinarity and Higher Education*. J. Kockelmans, 93-122. University Park: The Pennsylvania State University Press.

88. Fouke, J. and K. Brodie. 2003. *Symposium on catalyzing team science*. Bioengineering Consortium (BECON), National Institutes of Health.

89. Friedman, R. S., and R. C. Friedman. 1985. Organized Research Units of Academe Revisited. *Managing High Technology: An Interdisciplinary Perspective*. B. W. Mar, W. T. Newell, and B. O. Saxberg, 75-91.

90. Frost, Susan H. 2001. "Intellectual Initiatives: Working Across Disciplines, Schools and Institutions." *Emory: Provost Initiatives*, Emory University, Atlanta.

91. Garrett Russell, M. 1990. The Impact of Interdisciplinary Activities on Departmental Disciplines. *International Research Management*. P. H. Birnbaum-More, Frederick A. Rossini, and Donald R. Baldwin, 81-96. New York: Oxford University Press.

92. Gaughan, M., and S. Robin. 2004. National Science Training Policy and Early Scientific Careers in France and the United States. *Research Policy* 33:569-81.

93. Geneuth, J., I. Chompalov, and W. Shrum. 2000. How Experiments Begin: The Formation of Scientific Collaborations. *Minerva* 38, no. 3:311-48.

94. Gershon, D. 2000. Changing the Face of Training for Science at the Interface. *Nature* 404, no. 6775:315-6.

95. Gershon, D. 2000. Crossing the Divide Between Theory and Practice. *Nature* 404, no. 6775:316.

96. Gershon, D. 2000. Laying a Firm Foundation for Interdisciplinary Research Endeavours. *Nature* 406, no. 6791:107-8.

97. Gershon, D. 2000. Pushing the Frontiers of Interdisciplinary Research: an Idea Whose Time Has Come. *Nature* 404, no. 6775:313-5.

98. Goebel, G. "Microwave Radar and The MIT Rad Lab." Web page, [accessed 10 March 2004]. Available at http://www.vectorsite.net/ttwiz3.html.

99. Golde, C. M. and H. A. Gallagher. 1999. The Challenges of Conducting Interdisciplinary Research and Traditional Doctoral Programs. *Ecosystems* 2:281-85.

100. Gonzalez, C. 2003. The role of the graduate school in interdisciplinary programs: The University of California, Davis, Budget Model. *Communicator: Council of Graduate Schools* 26, no. 5:1-12.

101. Goulet, R. R. 1985. The National Science Foundation and High Tech Interdisciplinary Research. *Managing High Technology: An Interdisciplinary Approach.* W. T. Newell and B. O. Saxberg B.W. Mar, 179-83. New York: Elsevier.

102. Grace, D. J. and F. A. Rossini. 1990. Georgia Tech Research Institute: An Interdisciplinary Perspective. *International Research Management.* P. H. Birnbaum-More, Frederick A. Rossini, and Donald R. Baldwin, 105-13. New York: Oxford University Press.

103. Gregorian, V. 2004. Colleges Must Retain the Unity of Knowledge. *The Chronicle of Higher Education* 50, no. 39:1-5.

104. Guetzkow, J., and M. Lamont. 2002. Evaluating Creative Minds: The Assessment of Originality in Peer Review. *Creativity* July.

105. Gusdorf, G. 1977. Past, Present and Future in Interdisciplinary Research. *International Social Science Journal.* 29, no. 4:580-600.

106. Haberli, R., W. Grossenbacher-Mansuy, J. Thompson-Klein, A. Bill, R. Scholz, and M. Welti. 2001. Summary and Synthesis. *Transdisciplinarity: Joint Problem Solving among Science, Technology, and Society.* J. Thompson-Klein, W. Grossenbacher-Mansuy, R. Haberli, A. Bill, R.W. Scholz, and M. Welti, 3-23. Boston: Birkhauser Verlag.

107. Haggin, J. 1995. Illinois' Beckman Institute Targets Disciplinary Barriers to Collaboration. *Chemical and Engineering News* 73, no. 10:32-39.

108. Hagoel, L. 2002. Crossing Borders: Toward a Trans-Disciplinary Scientific Identity. *Studies in Higher Education* 27, no. 3:297-308.

109. Hansson, B. 1999. Interdisciplinarity: For what purpose? *Policy Sciences* 32:339-43.

110. Hattery, L. H. 1979. Interdisciplinary Research Management: Research Needs and Opportunities. *Interdisciplinary Research Groups: Their Management and Organization.* R. T Barth, and R. Steck, 9-25. IGRIP.

111. Hoag, H. 2003. Building Bridges. *Nature* 425:882-83.

112. Hollaender, K. 2001. Reflections on the Interactive Sessions: From Scepticism to Good Practices. *Transdisciplinarity: Joint Problem Solving among Science, Technology, and Society.* J. Thompson-Klein, W. Grossenbacher-Mansuy, R. Haberli, A. Bill, R.W. Scholz, and M. Welti Boston: Birkhauser-Verlag.

113. Hollingsworth, R., and E. J. Hollingsworth. 2000. Major Discoveries and Biomedical Research Organizations: Perspectives on Interdisciplinarity, Nurturing Leadership, and Integrated Structure and Cultures. *Practising Interdisciplinarity.* P. Weingart, and N. Stehr, 215-44. Toronto: University of Toronto Press.

114. Holton, G. 1998. *The Scientific Imagination.* Cambridge, MA: Harvard University Press.

115. Holton, G., and H. Jurkowitz E. Chang. 1996. How a Scientific Discovery is Made: A Case History. *American Scientist* 84, no. 4:364-75.

116. Hooper, J. W., and F. A. Rossini. 1990. The Development of an Interdisciplinary University-Based Microelectronics Research Center. *International Research Management.* Eds. P. H. Birnbaum, Frederick A. Rossini, and Donald R. Baldwin, 114-21. New York: Oxford University Press.

117. Horlick-Jones, T., and J. Sime. 2004. Living on the border: knowledge, risk and transdisciplinarity. *Futures* 36, no. 4:441-56.

118. Hursh, B., P. Haas, and M. Moore. 1983. An Interdisciplinary Model to Implement General Education. *Journal of Higher Education* 54:42-59.

119. Jeffrey, P. 2003. Smoothing the Waters: Observations on the Process of Cross-Disciplinary Research Collaboration. *Social Studies of Science* 33, no. 4:539-62.
120. Jewitt, G. P. W., and A. H. M. Görgens. 2000. Facilitation of interdisciplinary collaboration in research: lessons from a Kruger National Park Rivers Research Programme project. *South African Journal of Science* 96:411-14.
121. Kafatos, F. C., and T. Eisner. 2004. Unification in the Century of Biology. *Science* 303:1257.
122. Karlqvist, A. 1999. Going beyond disciplines: The meanings of interdisciplinarity. *Policy Sciences* 32:379-83.
123. Kates, R. W. 1989. The great questions of science and society do not fit neatly into single disciplines. *The Chronicle of Higher Education* 35:B2-B3.
124. Kessel, F., P. Rosenfield, and N. Anderson. 2003. "Expanding the Boundaries of Health and Social Science: Case Studies in Interdisciplinary Innovation." Oxford University Press, Oxford.
125. Klein, J. T. 1996. *Crossing Boundaries: Knowledge, Disciplinarities, and Interdisciplinarities.* Charlottesville: University Press of Virginia.
126. Klein, J. T. 1990. *Interdisciplinarity: History, Theory, & Practice.* Detroit: Wayne State University Press.
127. Klein, J. T. 2004. Prospects for Transdisciplinarity. *Futures* 36:515-26.
128. Klein, J. T., and W. H. Newell. 1997. Handbook of the Undergraduate Curriculum: A Comprehensive Guide to Purposes, Structures, Practices, and Change. *Advancing Interdisciplinary Studies.*, 393-415. San Francisco: Jossey-Bass.
129. Klein, J. T. Porter A. L. 1990. Preconditions for Interdisciplinary Research. *International Research Management: Studies in Interdisciplinary Methods from Business, Government, and Academia.* P. H. Birnbaum-More, F. A. Rossini, and D. R. Baldwin, 11-19. New York: Oxford University Press.
130. Kostoff, R. N. 2002. Overcoming Specialization. *BioScience* 52, no. 10:937-41.
131. Kostoff, R. N. 1997. Peer review: The appropriate GPRA metric for research. *Science* 277:651-52.
132. Kreeger, K. 2003. The Learning Curve. *Nature* 424:234-35.
133. Kusmierek, K. N. and M. Piontek. 2002. Content, Consciousness, and Colleagues: Emerging Themes from a Program Evaluation of Graduate Student Progress Toward Multidisciplinary Science. *42nd Annual Association for Institutional Research Forum.*
134. Landa, J. T. and M. T. Ghiselin. 1999. The emerging disciplines of bioeconomics: aims and scope of the Journal of Bioeconomics. *Journal of Bioeconomics* 1, no. 1:5-12.
135. Lattuca, L. R. 2001. *Creating Interdisciplinarity: Interdisciplinary Research and Teaching among College and University Faculty.* Nashville: Vanderbilt University Press.
136. Lawrence, R. J. 2004. Housing and health: from interdisciplinary principles to transdisciplinary research and practice. *Futures* 36, no. 4:487-502.
137. Lawrence, R. J. and C. Despres. 2004. Futures of Transdisciplinarity. *Futures* 36, no. 4:397-405.
138. Lind, I. 1999. Organizing for interdisciplinarity in Sweden: The case of *tema* at Linköping University. *Policy Sciences* 32:415-20.
139. Lindas, N. 1979. Conclusions from the American Society for Public Administration's Assessment of Four Interdisciplinary Research Management Projects. *Interdisciplinary Research Groups: Their Management and Organization.* R. T. Barth, and R. Steck, 278-94. IGRIP.
140. Lindbeck, A. "The Sveriges Riksbank (Bank of Sweden) Prize in Economic Sciences in Memory of Alfred Nobel 1969-2000." Web page, [accessed 27 February 2004]. Available at http://www.nobel.se/economics/articles/lindbeck/index.html.

141. Mansilla, V. B. and H. Gardner. 2003. Assessing Interdisciplinary Work at the Frontier. An Empirical Exploration of 'Symptoms of Quality." *Interdisciplines* http://www. interdisciplines.org/interdisciplinarity/papers/6.

142. Mar, B., W. T. Newell, and B. O. Saxberg. 1985. *Managing High Technology: An Interdisciplinary Perspective,* North Holland-Elsevier, Amsterdam.

143. Marshall, L. H., W. A. Rosenblith, P. Gloor, G. Krauthamer, C. Blakemore, and S. Cozzens. 1996. Early History of IBRO: The Birth of Organized Neuroscience. *Neuroscience* 72, no. 1:283-306.

144. McCain, K. W., and P. J. Whitney. 1994. Contrasting Assessments of Interdisciplinarity in Emerging Specialties: The Case of Neural Networks Research. *Knowledge: Creation, Diffusion, Utilization* 15, no. 3:285-306.

145. McCorcle, M. 1982. Critical Issues in the Functioning of Interdisciplinary Groups. *Small Group Behavior* 13:291-310.

146. McNeill, D. 1999. On Interdisciplinary Research: with particular reference to the field of environment and development. *Higher Education Quarterly* 53, no. 4:312-32.

147. Metzger, N. and R. N. Zare. 1999. Interdisciplinary Research: From Belief to Reality. *Science* 283, no. 5402:642.

148. Mey, H. 2001. Impact on Science Management and Science Policy. *Transdisciplinarity: Joint Probelm Solving among Science, Technology, and Society.* J. Thompson-Klein, W. Grossenbacher-Mansuy, R. Haberli, A. Bill, R.W. Scholz, and M. Welti, 253-59. Basel, Boston, Berlin: Birkhauser-Verlag.

149. Montgomery, S. 2004. Of Towers, Walls, and Fields: Perspectives on Language in Science. *Science* 303:1333-35.

150. Morgan, G. D., K. Kobus, K. K. Gerlach, C. Neighbors, C. Lerman, D. B. Abrams, and B. K. Rimer. 2003. Facilitating transdisciplinary research: The experience of the transdisciplinary tobacco use research centers. *Nicotine and Tobacco Research* 6, no. Suppl. 1:S11-S19.

151. Myers, G. 1993. Centering: Proposals for an Interdisciplinary Research center. *Science, Technology, & Human Values* 18, no. 4:433-59.

152. National Academy of Public Administration. 2004. "National Science Foundation: Governance and Management for the Future." 2015-001. Washington, D.C.

153. National Research Council. 1990. *Interdisciplinary Research: Promoting Collaboration Between the Life Sciences and Medicine and the Physical Sciences and Engineering.* Washington, D.C.: National Academy Press.

154. Nealson, K. and W. A. Ghiorse. 2000. Geobiology: Exploring the Interface between the Biosphere and the Geosphere. American Academy of Microbiology.

155. Nilles, J. M. 1975. Interdisciplinary Research Management in the University Environment. *Journal of the Society of Research Administrators* VI, no. 4:9-16.

156. O'Toole, J. C., G. H. Toenniessen, T. Murashige, R. R. Harris, and R. W. Herdt. The Rockefeller Foundation's International Program on Rice Biotechnology. *Proceedings of the Fouth International Rice Genetics Symposium* G. S. Khush, D. S. Brar, and B. Hardy.

157. Pain, E. 2003. Multidisciplinary Resources. *Science's Next Wave* http://nextwave. sciencemag.org/cgi/content/full/s002/12/30/16.

158. Palmer, C. 2001. "Work at the Boundaries of Science." *Work at the Boundaries of Science,* Dordrecht: Kluwer.

159. Palmer, C. L. 1999. Structures and Strategies of Interdisciplinary Science. *Journal of the American Society for Information Science.*

160. Palmer, C. L. and L. J. Neumann. 2002. The Information Work of Interdisciplinary Humanities Scholars: Exploration and Translation. *Library Quarterly* 72, no. 1:85-117.

161. Payne, S. L. 2001. Applying More Integrative Potenital for IDS Program Planning and Developemnt. *Issues in Integrative Studies* 19:149-69.
162. Pfirman, S. L., J. P. Collins, S. Lowes, and A. F. Michaels. in press. To Thrive and Prosper: Hiring, Supporting, and Tenuring Interdisciplinary Scholars.
163. Pinson, D. 2004. Urban planning: an 'undisciplined' discipline? *Futures* 36, no. 4:503-13.
164. Plater, W. M. 1995. Future work: Faculty time in the 21st century. *Change* 27, no. 3:23-33.
165. Porter, A. L. 1983. Interdisciplinary Research: Current Experience in Policy and Performance. *Interdisciplinary Science Reviews* 8, no. 2:158-67.
166. Porter, A. L. and D. E. Chubin. 1985. An Indicator of Cross-Disciplinary Research. *Scientometrics* 8, no. 3-4:161-76.
167. Porter, A. L., and F. A. Rossini. 1985. Forty Interdisciplinary Research Projects: Multiple Skills and Peer Review. *Managing High Technology: An Interdisciplinary Perspective.* B. W. Mar, W. T. Newell, and B. O. Saxberg, 103-12. New York: Elsevier.
168. Porter, A. L., and F. A. Rossini. 1984. Interdisciplinary research redefined: multi-skill problem-focussed research in the STRAP framework. *R&D Management* 14, no. 2:105-11.
169. Porter, A. L., and F. A. Rossini. 1986. Multiskill Research. *Knowledge: Creating, Diffusion, Utilization* 7, no. 3:219-46.
170. Porter, A. L., and F. A. Rossini. 1985. Peer Review of Interdisciplinary Research Proposals. *Science Technology & Human Values* 10, no. 3:33-38.
171. Pray, L. A. 2003. Interdisciplinarity in science and engineering: Academia in transition. *Science's Next Wave* http://nextwave.sciencemag.org/cgi/content/full/2003/01/15/5?
172. Qin, J., F. W Lancaster, and B. Allen. 1997. Types and Levels of Collaboration in Interdisciplinary Research in the Sciences. *Journal of the American Society for Information Science* 48, no. 10:893-916.
173. Ramadier, T. 2004. Transdisciplinarity and its challenges: the case of urban studies. *Futures* 36, no. 4:423-39.
174. Rhoten, D. 2004. Interdisciplinary Research: Trend or Transition. *Items and Issues* 5, no. (1-2):6-11.
175. Rhoten, D. Final Report, National Science Foundation BCS-0129573: A Multi-Method Analysis of the Social and Technical Conditions for Interdisciplinary Collaboration. September 29, 2003. Available at: http://www.hybridvigor.net/interdis/pubs/hv_pub_interdis-2003.09.29.pdf.
176. Rhoten, D. 2002. Organizing change from the inside out: Emerging models of intra-organizational collabotation in philanthoropy. *The Hybrid Vigor Institute* http://www.hybridvigor.net/interdis/pubs/hv_pub_interdis-2002.10.30.pdf.
177. Rinia, E. J, Th. N. Van Leeuwen, H. G. van Vuren, and A. F. J van Raan. 2001. Influence of interdisciplinarity on peer-review ad bibliometric evaluations in physics research. *Research Policy* 30:357-61.
178. Roberston, D. W., D. K. Martin, and P. A. Singer. 2003. Interdisciplinary Reseach: Putting the methods under the microscope. *Medical Research Methodology* 3, no. 20.
179. Roberts, J. A., and R. E. Barnhill. 2001. Engineering Togetherness: An Incentive System for Interdisciplinary Research. *ASEE/IEEE Frontiers in Education Conference.*
180. Rogers, J. 2000. The intellectual consequences of the research assessment exercise: A Response. *History of the Human Sciences* 13, no. 2:101-6.
181. Rogers, J. D. and B. Bozeman. 2002. A churn model of scientific knowledge value: Internet researchers as a knowledge value collective. *Research Policy* 31, no. 5:769-94.

182. Rossini, F., A. L. Porter, D. E. Chubin, and T. Connolly. 1984. Cross-Disciplinarity in the Biomedical Sciences: A Preliminary Analysis of Anatomy. *Managing Interdisciplinary Research*. S. R. Epton, R. L. Payne, and A. W. Pearson, 176-84. Chichester: Wiley.

183. Rossini, F. A. and A. L. Porter. 1979. Frameworks for integrating interdisciplinary research. *Research Policy* 8:70-79.

184. Russell, M. G. 1982. *Enabling Interdisciplinary Research: Perspectives from Agriculture, Forestry, and Home Economics*, Miscellaneous Publication #19. Agricultural Experiment Station, University of Minnesota, St. Paul, MN.

185. Russell, M. G. 1985. Peer Review and Interdisciplinary Research: Flexibility and Responsiveness. *Managing Interdisciplinary Research*. S. R. Epton, R. L. Payne, and A. W. Pearson, 184-202. New York: John Wiley and Sons.

186. Russell, M. G. and R. J. Sauer. 1983. Creating Administrative Environments for Interdisciplinary Research. *Journal of the Society of Research Administrators* 14, no. 4:21-31.

187. Sackett, W. T. 1990. Interdisciplinary Research in a High-Technology Company. *International Research Management*. P. H. Birnbaum-More, Frederick A. Rossini, and Donald R. Baldwin, 60-72. New York: Oxford University Press.

188. Sanders, C. 2003. Breaking Down the Borders of Science. *New York Times* 1570:8.

189. Saxberg, B. O., W. T. Newell, and B. W. Mar. 1981. Interdisciplinary Research: A Dilemma for University Central Administration. *Journal of the Society of Research Administrators* 13:25-43.

190. Schneidewind, U. 2001. Mobilizing the Intellectual Capital of Universities. *Transdisciplinarity: Joint Problem Solving among Science, Technology, and Society*. J. Thompson-Klein, W. Grossenbacher-Mansuy, R. Haberli, A. Bill, R.W. Scholz, and M. Welti, 94-100. Boston: Birkhauser-Verlag.

191. Sewell, W. H. 1989. Some Reflections on the Golden Age of Interdisciplinary Social Psychology. *Social Psychology Quarterly* 52, no. 2:88-97.

192. Shanken, E. A. "Bibliography of Interdisciplinary Collaboration." Web page, [accessed 17 March 2004]. Available at *http://www.artexetra.com/biblio_interdisciplinary.html*.

193. Simonton, D. K. 2004. *Creativity in Science: Chance, Logic, Genius, and Zeitgeist*. New York: Cambridge University Press.

194. Simonton, D. K. 2003. Scientific Creativity as Constrained Stochastic Behavior: The Integration of Product, Person, and Process Perspectives. *Psychological Bulletin* 129, no. 4:475-94.

195. Sjolander, S. 1985. "Long-term and Short-term Interdisciplinarity." *Inter-Disciplinarity Revisited*, L. Levin, and I. Eds. Stockholm: Lind.

196. Smith, R., M. T. Spina, J. Grayson, and M. L. Zuzolo. 1997. Research Administration Through a Decade of Change. *SRA Journal* 29, no. 1,2:37-44.

197. Smoliner, C., R. Haberli, and M. Welti. 2001. Mainstreaming Transdisiplinarity: A Research-Political Campaign. *Transdisciplinarity: Joint Problem Solving among Science, Technology, and Society*. J. Thompson-Klein, W. Grossenbacher-Mansuy, R. Haberli, A. Bill, R.W. Scholz, and M. Welti, 263-71. Boston: Birkhauser-Verlag.

198. Spiegel-Rosing, I. and D. de Solla Price. 1977. *Science, Technology, and Society: A Cross-Disciplinary Perspective*. Beverly Hills: Sage Publications.

199. Sproull, R., and H. Hall. 1987. " Multidisciplinary Research and Education Programs in Universities." Washington, D.C.: National Academy Press.

200. Steele, T. W. and J. C. Stier. 2000. The Impact of Interdisciplinary Research in the Environmental Sciences: A Forestry Case Study. *Journal of the American Society for Information Science* 51, no. 5:476-84.

201. Steinheider, B. and G. Legrady. 2004. Interdisciplinary Collaboration in Digital Media Arts: A Psychological Perspective on the Production Process. *Leonardo* 37, no. 4: http://mitpress2.mit.edu/e-journals/Leonardo/isast/journal/toc374.html.

202. Stember, Marilyn. 1991. Advancing the social sciences through the interdisciplinary enterprise. *Social Science Journal* 28, no. 1:1, 14.

203. Stokols, D., J. Fuqua, J. Gress, R. Harvey, K. Phillips, L. Baezconde-Garbanati, J. Unger, P. Palmer, M. A. Clark, S. M. Colby, G. Morgan, and W. Trochim. 2003. Evaluating transdisciplinary science. *Nicotine and Tobacco Research* 6, no. Suppl. 1:S21-S39.

204. Stone, A. R. 1969. The Interdisciplinary Research Team. *Journal of Applied Behavioral Sciences* 5, no. 3:351-65.

205. Straus, R. 1973. Departments and Disciplines: Stasis and Change. *Science* 182, no. 4115:895-98.

206. Submcommittee on Research Business Models. 2003. Alignment of Funding Mechanisms with Scientific Opportunities. *Regional Forum on Research Business Models* White House Office of Science and Technology Policy (OSTP) National Science and Technology Council (NSTC).

207. Suh, N. P. 1987. The ERCs: What Have We Learned. *Engineering Education* October: 16-18.

208. Suk, W. A., B. E. Anderson, C. L. Thompson, D. A. Bennett, and D. C. Vandermeer. 1999. Creating Multidisciplinary Research Opportunities. *Environmental Science & Technology* 33, no. 11:241A-4A.

209. Swoboda, W. W. 1979. Disciplines and Interdisciplinarity: A Historical Perspectives. *Interdisciplinarity and Higher Education*. Joseph Kockelmans, 49-84. University Park: The Pennsylvania State University Press.

210. Teagarden, M. B. 1998. Unbundling the intellectual joint venture process. The case of multinational, multifunctional interdisciplinary research consortia. *Journal of Managerial Psychology* 13, no. 3,4:178-87.

211. Trouillot, M-R. 1996. Open the Social Sciences. *Crosscurrents* 3, no. 2.

212. University of Michigan. 2001. *New Openings for the Research University: Advancing Collaborative, Integrative, and Interdisciplinary Research and Learning: University of Michigan Special Emphasis Self Study*, University of Michigan.

213. University of Washington. 2002. *Self-Study: Program on the Environment*, University of Washington.

214. Van Dusseldorp, S. and D. Wigboldus. 1994. Interdisciplinary Research for Integrated Rural Development in Developing Countries. *Issues in Integrative Studies* 12:93-138.

215. Waintraub, J. L. 1997. Institution wide reform through NSF supported projects. *ASEE Annual Conference Proceedings*.

216. Walsh, S., and T. Bartlett. 2003. U. of California-Merced Takes Interdisciplinary Approach; Virginia Tech Sends a Mixed Message. *The Chronicle of Higher Education* 49, no. 28:A10.

217. Weingart, P., and N. Stehr. 2000. *Practising Interdisciplinarity*. Toronto: University of Toronto Press.

218. Whitley, R. 1982. The rise and decline of university disciplines in the sciences. *Issues in Interdisciplinarity*. J. H. P. Paelinck, 17-43. Rotterdam: Erasmus University Press.